Embedded Artificial Intelligence
Devices, Embedded Systems, and Industrial Applications

RIVER PUBLISHERS SERIES IN COMMUNICATIONS AND NETWORKING

Series Editors:

ABBAS JAMALIPOUR
The University of Sydney
Australia

MARINA RUGGIERI
University of Rome Tor Vergata
Italy

The "River Publishers Series in Communications and Networking" is a series of comprehensive academic and professional books which focus on communication and network systems. Topics range from the theory and use of systems involving all terminals, computers, and information processors to wired and wireless networks and network layouts, protocols, architectures, and implementations. Also covered are developments stemming from new market demands in systems, products, and technologies such as personal communications services, multimedia systems, enterprise networks, and optical communications.

The series includes research monographs, edited volumes, handbooks and textbooks, providing professionals, researchers, educators, and advanced students in the field with an invaluable insight into the latest research and developments.

Topics included in this series include:-

- Communication theory
- Multimedia systems
- Network architecture
- Optical communications
- Personal communication services
- Telecoms networks
- Wifi network protocols

For a list of other books in this series, visit www.riverpublishers.com

Embedded Artificial Intelligence
Devices, Embedded Systems, and Industrial Applications

Editors

Ovidiu Vermesan

SINTEF, Norway

Mario Diaz Nava

STMicroelectronics, France

Björn Debaillie

imec, Belgium

Routledge
Taylor & Francis Group
LONDON AND NEW YORK

Published 2023 by River Publishers
River Publishers
Alsbjergvej 10, 9260 Gistrup, Denmark
www.riverpublishers.com

Distributed exclusively by Routledge
4 Park Square, Milton Park, Abingdon, Oxon OX14 4RN
605 Third Avenue, New York, NY 10017, USA

Embedded Artificial Intelligence / by Ovidiu Vermesan, Mario Diaz Nava, Björn Debaillie.

Routledge is an imprint of the Taylor & Francis Group, an informa business

ISBN 978-87-7022-821-3 (print)
ISBN 978-10-0088-191-2 (online)
ISBN 978-1-003-39444-0 (ebook master)

While every effort is made to provide dependable information, the publisher, authors, and editors cannot be held responsible for any errors or omissions.

Dedication

"The question is not whether intelligent machines can have any emotions, but whether machines can be intelligent without any emotions."

– Marvin Minsky

"Our ultimate objective is to make programs that learn from their experience as effectively as humans do. We shall... say that a program has common sense if it automatically deduces for itself a sufficiently wide class of immediate consequences of anything it is told and what it already knows."

– John McCarthy

"It is customary to offer a grain of comfort, in the form of a statement that some peculiarly human characteristic could never be imitated by a machine. I cannot offer any such comfort, for I believe that no such bounds can be set."

– Alan Turing

Acknowledgement

The editors would like to thank all the contributors for their support in the planning and preparation of this book. The recommendations and opinions expressed in the book are those of the editors, authors, and contributors and do not necessarily represent those of any organizations, employers, or companies.

Ovidiu Vermesan
Mario Diaz Nava
Björn Debaillie

Contents

Preface

Embedded Artificial Intelligence

Embedded edge artificial intelligence (AI) reduces latency, increases the speed of processing tasks, and reduces bandwidth requirements by reducing the among of data transmitted, and costs by introducing cost-effective and efficient low power hardware solutions allowing processing data locally. New embedded AI techniques offer high data security, decreasing the risks to sensitive and confidential data and increasing the dependability of autonomous technologies.

Embedded edge devices are becoming more and more complex, heterogeneous, and powerful as they incorporate a combination of hardware components like central processing units (CPUs), microcontroller processing units (MCUs), graphics processing units (GPUs), digital signal processors (DSPs), image signal processors (ISPs), neural processing units (NPUs), field-programmable gate arrays (FPGAs), application specific integrated circuits (ASICs) and other accelerators to perform multiple forms of machine learning (ML), deep learning (DL) and spiking neural network (SNN) algorithms. Embedded edge devices with dedicated accelerators can perform matrix multiplication significantly faster than CPUs, and ML/DL algorithms implemented in AI frameworks and edge AI platforms can efficiently exploit these hardware components.

Processing pipelines, toolchains, and flexible edge AI software architectures can provide specific system-on-a-chip (SoC), system-on-module (SoM) and application types for optimised run-time support. These tools can facilitate the full exploitation of heterogeneous SoC/SoM capabilities for ML/DL and maximise component reuse at the edge.

The book offers complete coverage of the topics presented at the International Workshop on Embedded Artificial Intelligence (EAI) - Devices, Systems, and Industrial Applications" in Milan, Italy 19 September 2022, as part of the ESSCIRC/ESSDERC 2022 European Solid-state Circuits and Devices Conference held in Milan, Italy, combining ideas and concepts developed by researchers and practitioners working on creating edge AI methods, techniques, and tools for industrial applications.

The book explores the challenges faced by AI technologies embedded into electronic systems and applied to various industrial sectors by highlighting essential topics, such as embedded AI for semiconductor manufacturing; trustworthiness, verification, validation and benchmarking of AI systems and technologies; the design of novel AI-based hardware architectures; neuromorphic implementations; edge AI platforms; and AI-based workflows deployments on hardware.

This book is a valuable resource for researchers, post-graduate students, practitioners and technology developers interested in gaining insight into embedded AI, ML, DL, SNN and the technology trends advancing intelligent processing at the edge. It covers several embedded AI research topics and is structured into ten articles. A brief introduction of each article is discussed in the following paragraphs.

Ana Pinzari, Thomas Baumela, Liliana Andrade, Marcello Copolla and Frédéric Pétrot: "Power Optimised Wafermap Classification for Semiconductor Process Monitoring" introduce a power efficient neural network architecture specifically designed for embedded system boards that includes microcontroller and edge tensor processing units. Experiments show that the analysis of the control of wafers can be achieved in real-time with an accuracy of 99.9% (float) and 97.3% (8-bit integer) using less than 2W.

Roland Müller, Bijoy Kundu, Elmar Herzer, Claudia Schuhmann and Loreto Mateu: "Low-Power Analog In-memory Computing Neuromorphic Circuits" present the ASIC design and validation results of a neuromorphic circuits comprising synaptic weights and neurons. This design includes batch normalization, activation function, and offset cancelation circuits. The ASIC shows excellent results: 12 nJ per inference with 5μs latency.

Loreto Mateu, Johannes Leugering, Roland Müller, Yogesh Patil, Maen Mallah, Marco Breiling and Ferdinand Pscheidl: "Tools and Methodologies for Edge-AI Mixed-Signal Inference Accelerators" present how a toolchain to facilitate design, training, and deployment of artificial neural networks in dedicated hardware accelerators allows to optimize and verify the hardware design, reach the targeted KPIs, and reduce the time-to-market.

Sourav De, Sunanda Thunder, David Lehninger, Michael P.M. Jank, Maximilian Lederer, Yannick Raffel, Konrad Seidel, and Thomas Kämpfe: "Low-Power Vertically Stacked One Time Programmable Multi-bit IGZO-Based BEOL Compatible Ferroelectric TFT Memory Devices with Lifelong Retention for Monolithic 3D-Inference Engine Applications" discuss and demonstrate an IGZO-based one-time programmable FeFET memory device with multilevel coding and lifelong retention capability. The synaptic device shows to achieve 97% for inference-only application with MNIST data and an accuracy degradation of only 1.5% over 10 years. The proposed inference engine also showed superior energy efficiency and cell area.

Bernhard Lippmann, Matthias Ludwig, and Horst Gieser: "Generating Trust in Hardware through Physical Inspection" address the image processing methods for physical inspection within the semiconductor manufacturing process and physical layout to provide trustworthiness in the produced microelectronics hardware. The results are presented for a 28nm process including a proposed quantitative trust evaluation scheme based on feature similarities.

Ivan Miro-Panades, Inna Kucher, Vincent Lorrain, and Alexandre Valentian: "Meeting the Latency and Energy Constraints on Timing-critical Edge-AI Systems" explore a novel architectural approach to overcome such limitations by using the attention mechanism of the human brain. The energy-efficient design includes a small NN topology (i.e., MobileNet-V1) to be completely integrable on-chip; heavily quantized (4b) weights and activations, and fixed bio-inspired extraction layers in order to limit the embedded memory capacity to 600kB.

Hannah Bos and Dylan Muir: "Sub-mW Neuromorphic SNN Audio Processing Applications with Rockpool and Xylo" apply a new SNN architecture designed for temporal signal processing, using a pyramid of synaptic time constants to extract signal features at a range of temporal scales. The architecture was demonstrated on an ambient audio classification task, deployed to the Xylo SNN inference processor in streaming mode. The application achieves high accuracy (98 %) and low latency (100 ms) at low power ($<100\mu$W dynamic inference power).

Jianyu Zhao, Cecilia Carbonelli and Wolfgang Furtner: "An Embedding Workflow for Tiny Neural Networks on ARM Cortex-M0(+) Cores" provide a description and propose an end-to-end embedding workflow focused on tiny neural network deployment on Arm® Cortex®-M0(+) cores. With this, up to 73.9% of the memory footprint could be reduced. While reducing the manual effort of network embedding to the minimum, the workflow remains flexible enough to allow for customizable bit shifts and different layer combinations.

Ovidiu Vermesan and Marcello Copolla: "Edge AI Platforms for Predictive Maintenance in Industrial Applications" provide an assessment and comparative analysis of several existing edge AI platforms and workflows including some of the most essential architectural elements of differentiation

(AEDs) in edge AI-based industrial applications, such as analytic capabilities in the time and frequency domains, features visualisation and exploration, microcontroller (Arm® Cortex®-M cores) emulator and live tests, support for ML, DL, and using ML core capabilities implemented in the sensors.

Rameez Ismail and Zhaorui Yuan: "Food Ingredients Recognition Through Multi-label Learning" describe deep multi-label learning approaches and related models to detect an arbitrary number of ingredients in a dish image. With an average precision score of 78.4% using a challenging dataset (Nutrition5K), this approach forms a strong baseline for future exploration.

Editors Biography

Ovidiu Vermesan holds a PhD degree in microelectronics and a Master of International Business (MIB) degree. He is Chief Scientist at SINTEF Digital, Oslo, Norway. His research interests are in smart systems integration, mixed-signal embedded electronics, analogue neural networks, edge artificial intelligence and cognitive communication systems. Dr. Vermesan received SINTEF's 2003 award for research excellence for his work on the implementation of a biometric sensor system. He is currently working on projects addressing nanoelectronics, integrated sensor/actuator systems, communication, cyber–physical systems (CPSs) and Industrial Internet of Things (IIoT), with applications in green mobility, energy, autonomous systems, and smart cities. He has authored or co-authored over 100 technical articles, conference/workshop papers and holds several patents. He is actively involved in the activities of European partnership for Key Digital Technologies (KDT). He has coordinated and managed various national, EU and other international projects related to smart sensor systems, integrated electronics, electromobility and intelligent autonomous systems such as E^3Car, POLLUX, CASTOR, IoE, MIRANDELA, IoF2020, AUTOPILOT, AutoDrive, ArchitectECA2030, AI4DI, AI4CSM. Dr. Vermesan actively participates in national, Horizon Europe and other international initiatives by coordinating the technical activities and managing the various projects. He is the coordinator of the IoT European Research Cluster (IERC) and a member of the board of the Alliance for Internet of Things Innovation (AIOTI). He is currently the technical co-coordinator of the Artificial Intelligence for Digitising Industry (AI4DI) project.

Mario Diaz Nava has a Ph.D, and M.S. both in computer science, from Institut National Polytechnique de Grenoble, France, and B.S. in communications and electronics engineering from Instituto Politecnico National, Mexico. He has worked in STMicroelectronics since 1990. He has occupied different positions (Designer, Architect, Design Manager, Project Leader, Program Manager) in various STMicroelectronics research and development organisations. His selected project experience is related to the specifications and design of communication circuits (ATM, VDSL, Ultra-wideband), digital and analogue design methodologies, system architecture and program management. He currently has the position of ST Grenoble R&D Cooperative Programs Manager, and he has actively participated, for the last five years, in several H2020 IoT projects (ACTIVATE, IoF2020, Brain-IoT), working in key areas such as Security and Privacy, Smart Farming, IoT System modelling, and edge computing. He is currently leading the ANDANTE project devoted to developing neuromorphic ASICS for efficient AI/ML solutions at the edge. He has published more than 35 articles in these areas. He is currently a member of the Technical Expert Group of the PENTA/Xecs European Eureka cluster and a Chapter chair member of the ECSEL/KDT Strategic Research Innovation Agenda. He is an IEEE member. He participated in the standardisation of several communication technologies in the ATM Forum, ETSI, ANSI and ITU-T standardisation bodies.

Björn Debaillie leads imec's collaborative R&D activities on cutting-edge IoT technologies in imec. As program manager, he is responsible for the operational management across programs and projects, and focusses on strategic collaborations and partnerships, innovation management, and public funding policies. As chief of staff, he is responsible for executive finance and operations management and transformations. Björn coordinates semiconductor-oriented public funded projects and seeds new initiatives on high-speed communications and neuromorphic sensing. He currently leads the 35M€ TEMPO project on neuromorphic hardware technologies, enabling low-power chips for computation-intensive AI applications (www.tempo-ecsel.eu). Björn holds patents and authored international papers published in various journals and conference proceedings. He also received several awards, was elected as IEEE Senior Member and is acting in a wide range of expert boards, technical program committees, and scientific/strategic think tanks.

List of Figures

List of Tables

Power Optimized Wafermap Classification for Semiconductor Process Monitoring

Ana Pinzari, Thomas Baumela, Liliana Andrade, Marcello Coppola, and Frédéric Pétrot

Abstract—Today, the exploitation of AI solutions is very immersive and has wide applicability in virtually all industrial fields. In many sectors, the quality of the final product is the key to profitability. For the semi-conductor industry, this translates into production yield, the ratio of functional dies over the total number of dies produced. The control of wafer fabrication in the semiconductor industry is a fundamental task to ensure high yield. Analysis of the distribution of non-functional dies on a wafer is a necessary step to identify process drifts leading to their root causes. Current approaches use large-scale state-of-the-art neural networks running on GPUs to perform this analysis. Aiming at power efficiency, we propose a neural network architecture specifically designed to target embedded devices such as STMicroelectronics MP1 board or Google's Coral board that includes an edge tensor processing unit. Experiments show that we achieve this analysis in real-time with an accuracy of 99.9% (float) and 97.3% (8-bit integer) using less than 2W.

Index Terms—Process Control, Wafermap Classification, Deep Learning, Convolutional Neural Network Optimization, Hyper-parameter Tuning, Low-Energy Consumption, Power Reduction.

I. INTRODUCTION

YIELD is paramount in semiconductor process manufacturing, as it determines the financial viability of a production line for a given process. Since the start of the microelectronic VLSI industry, yield control has been the focus of the process engineers [1], and although the technology has matured in a tremendous manner, today's challenges are equally difficult to take up [2].

One important step in yield control is wafer testing (also known as Circuit Probe). A specialized test equipment is used to test each die on the wafer, and indicates which dies are functional and which are not. This is used to build a *binary wafermap* [3], i.e., 2-dimension image in which each pixel represents a die. A white pixel indicates a functional die, while a black

This work was conducted under the framework of the ECSEL AI4DI "Artificial Intelligence for Digitising Industry" project. The project has received funding from the ECSEL Joint Undertaking (JU) under grant agreement No 826060. The JU receives support from the European Union's Horizon 2020 research. *(Corresponding author: A. Pinzari).*

A. Pinzari is with Institute of Engineering Univ. Grenoble Alpes, France (e-mail: Ana.Pinzari@univ-grenoble-alpes.fr).

T. Baumela is with Institute of Engineering Univ. Grenoble Alpes, France (e-mail: Thomas.Baumela@univ-grenoble-alpes.fr).

L. Andrade is with Institute of Engineering Univ. Grenoble Alpes, France (e-mail: Liliana.Andrade@univ-grenoble-alpes.fr).

M. Coppola is with ST Microelectronics, Grenoble, 38000 France. (Marcello.Coppola@st.com).

F. Pétrot is with Institute of Engineering Univ. Grenoble Alpes, France (Frederic.Petrot@univ-grenoble-alpes.fr).

one indicates a non-functional die. The distribution of the black pixels is attached to a class of classical issues, which gives leads to the process engineer to identify the causes of the malfunctions.

Given the ability of recent neural networks to accurately perform classification, and the importance of wafermap classification in the semiconductor fabrication process, a lot of research has been done on the subject recently. Automating this process makes sense: human classification of abstract dots distributions is feasible for a small number of classes, but very complex when the number of classes is high. The existing approaches are using increasingly complex state-of-the-art network architectures, to gain a few tens of percent in accuracy. Given the low throughput of the test equipment, we believe that searching for a much smaller ad-hoc architecture that will be able to run on a micro-controller at production line speed will lead to much better power efficiency while only marginally degrading accuracy.

In this paper we present our low energy wafermap classification approach, put to work on actual foundry data from STMicroelectronics 28 nm fabrication facilities, which uses a purposely defined neural network, and makes use of advanced quantization techniques to work on a limited number of bits. This has the nice property of both limiting the memory used for parameter storage and minimizing the complexity of the operators used for computation.

This paper reviews the related works in Section II, introduces a brief description of the dataset used for training and classification in Section III, presents the main contributions related to the proposed neural network architecture in Section IV, describes the experiments and implementation performed on different embedded

devices in Section V, and concludes in Section VI.

II. RELATED WORKS

Although statistical approaches have been used for long to classify wafermaps in high-volume production [4], it is only recently that the used of deep neural network has been proposed to that aim. This is due to two factors: First the breakthrough brought by AlexNet [5] making neural networks clearly superior to any other algorithm for classification, and second the availability of an open wafermap dataset (WM-811K) donated to the community by a major foundry.

Wu *et al* [6] introduced the approach and made the dataset public, which led to a very large number of papers being published around this dataset, far too many to cite here. They all share one characteristic: the use of the most recent neural network architecture of the time, generally needing more computations and more parameters than the previous one, to gain little in accuracy. The WM-811K dataset is quite heterogeneous in size, shape, number of examples per classes, and has a small, 9, number of classes, whereas industrial practice is more in the order of 50, and class labelling is sometimes not correct. So, each published work does its own pre-processing to resize wafermaps and carries out data augmentation to even the class cardinality, making fair comparisons difficult.

In the next paragraphs we give an overview on recent research. In this context, [7] introduces a CNN model that can be used to classify patterns and to identify the cause of defects by using an image retrieval technique. To avoid the use of imbalanced datasets, the authors generate synthetic wafermaps modelled using a Poisson distribution. [8]

uses a CNN to automatically extract features and identify defects, in addition to use batch normalization and spatial dropout techniques to improve classification performances. During data pre-processing they use random rotations, horizontal flipping, width, and height shifts, shearing range, channel shifting and zooming as data augmentation techniques. The final dataset used by authors have 90K wafer defect images and it is fully balanced with 10K images per class. [9] proposes a new classification system, that based on active learning of CNN, allows the selection of a reduced and representative subset of unlabelled wafermaps that will be inspected by experts. Based on a model of four steps a small LeNet-5-like CNN architecture is trained using the initial labelled dataset, an uncertainty prediction in the unlabelled wafermaps is calculated, and using top-K selection methods, a new set of wafermaps is extracted to be manually inspected and merged with the original dataset. [10] also uses a CNN to detect defects but with the particularity that wafermaps are augmented using rotations techniques and converted in 2D arrays before training. This approach avoids the problems related to wafer size variations. The authors also apply data augmentation before training to take in account possible rotations related to the input wafermaps. [11] proposes to use a CNN encoder-decoder for data-augmentation and a depth-wise separable convolution for classification. The greatest advantage of this approach is that proposed model reduces the number of parameters by 30% and the amount of calculation by 75% in the WM-811K dataset. [12] proposes a neural network pre-training method based on self-supervised learning to improve classification performance on imbalanced datasets. A CNN-encoder is responsible to learn features by mapping similar wafermaps in a same space. Authors argue that self-supervised training methods improve classification performance when datasets with several unlabelled data are used, which is the case of the WM-811K dataset. [13] presents experimentations with simplified AlexNet, MobileNetV1, and VGG, so as to limit the number of parameters they require. This latter architecture leads to the best accuracy while requiring the least parameters.

Overall, among these works and others, only [11] evokes power efficiency consideration in a short paragraph, simply saying that the network fits into a NVidia Jetson Nano board (≈ 10 W) and performs a 5 frames per second inference on 64x64 images. None refers to power reduction techniques or power/accuracy trade-offs for neural network inference.

III. DESCRIPTION OF THE DATASET

Although identified long ago, the quality of data is a too often neglected issue when dealing with neural networks [14]. First, the number of elements must be large enough for the complexity of the problem, typically a few thousands for simple problems to a few millions for more complex ones. These elements must also be well balanced between the classes and labelled with care.

After a broad analysis of possible failures during the manufacturing process, the process engineers have defined 58 possible wafermap failure patterns. The wafers have a notch that is used to precisely align them within the machine during fabrication. This makes the orientation known, and allows to consider differently e.g., vertical and horizontal patterns. Each class contains about 2,200 images, resulting in a complete and well-balanced data set of 121,550 images.

Although build on purpose for data-confidentiality reasons, Figure 1 shows six wafermaps representative of 6 categories among the 58.

The binary wafermap images have an original resolution of 401×401. As the size of the input images has a major impact on the size of the network, it is very useful to resize them to reduce the number of parameters and computations. This process is worthwhile if the loss of information is minimal. A failure category has its own characteristics, so resizing must be done in a way that respects the patterns of the category to which an image belongs. After several experiments using state-of-the-art network architectures, we determined the target size to be 224×224 pixels. As a resizing method, we applied the nearest neighbour interpolation algorithm [15], which produces grayscale images. We then binarize the images simply by considering all non-white pixels as black and all white pixels as white. We observed that with this strategy there was no loss of information, and the image retains its own failure characteristics. With such reduced size black and white images, the memory space necessary to encode the image is minimized, and the even more interesting, the inputs to the neural network are single bits.

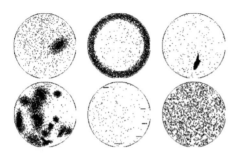

Fig. 1. Synthetic examples of wafermap failure patterns. From left to right and top to bottom, the big-cluster, wide-dense-edge-donut, fingerprint, complete-wafer, horizontal-dots-lines, and matrix classes are shown.

Now that we have explained our pre-processing steps, we can perform the neural network architecture search step.

IV. AD-HOC NEURAL NETWORK

Since our hardware targets are small, embedded devices in the watt power range, we must minimize the number of multiplication-accumulation operations, the size of the operands of these operations, and the number and size of the network parameters. We have two assets for that. First, our classification problem is somehow easy, compared to the current CIFAR100 or ImageNet challenges: we have black and white bitmaps instead of 256 or even true-colour pictures, and the number of classes is large but not huge. Second, much recent efforts have been put into getting rid of large floating-point representations for the data and parameters.

A. Neural Network Architecture Definition

Finding the appropriate NN architecture for a given problem is not a well formalized problem. This can be automated using AutoML [16] or Neural Architecture Search [17] approaches, but this requires an enormous number of resources, and is beyond the scope of our work. So, we classically got inspired by the existing CNNs approaches and following intuition, we made a few well-chosen experiments to converge towards a suitable architecture. We noticed that some defaults on the wafermaps are similar but have different scales, which is something the inception layers, as introduced in GoogLeNet [18], can deal with well. As our goal is to limit the number of layers, and to minimize the number of parameters while keeping a high prediction accuracy, we shall introduce only very few of them given their complexity.

In the end, we introduced a single inception layer within our architecture, which significantly improved the final accuracy.

Our final network architecture is depicted in Figure 2. Our model has a bit less than 0.5 million parameters, to be compared with the 58 million of AlexNet and the 6 million of GoogLeNet.

There are 17 network layers (convolution, subsampling and fully connected layers) of which 10 are learnable layers. The input shape is a $224\times224\times1$ bit image which is subsampled across the entire architecture into a $7\times7\times116$-dimensional vector.

Table I summarizes the description of our architecture. The model starts with a 7×7 convolution layer, followed by a max-pooling layer and an inception block. The first convolution has no padding, which allows the feature detector to work only on the pixels of the image. This upper part of the diagram extracts the most important features of the image. Next, the pooling layer maximizes the response of each feature map and reduces the space exploration of feature maps. The bottleneck layer (i.e., 1×1 convolution) reduces the number of parameters, and the last two consecutives 3×3 convolutional layers take the role to increase the feature maps volume again, before being flattened and passed to the classification function.

TABLE I
DESCRIPTION OF THE MODEL ARCHITECTURE
(K (N × N) MEANS K KERNELS OF SIZE N × N)

Conv2D	32 (7×7), stride of 2, no padding
MaxPool2D	(2×2), stride of 2
Inception Block	32 (1×1), 8 (1×1), 8 (1×1), MaxPool (3×3)
	32 (3×3), 32 (5×5), 32 (1×1)
MaxPool2D	(3×3), stride of 2
Conv2D	12 (1×1)
Conv2D	116 (3×3), stride of 2, padding with zeros
Conv2D	116 (3×3), stride of 2, padding with zeros
Dense/Softmax	58
Nr of parameters FLOPs	478,150 125,518,940

The intermediate layers use *Relu* as activation function and the last fully connected layer uses a *Softmax* activation to produce a probability score over the 58 classes (the sum of all probabilities must equal 1). Since we are facing a multi-class classification problem, *categorical_crosentroppy* is used as a feedback signal, and the weights adjustments during the backpropagation steps are carried out by the Adam optimizer. During the learning process an important aspect is the choice of hyperparameters. By varying the *batch size* and *learning rate*, the accuracy is remarkably improved. The *regularization* of these hyper-parameters

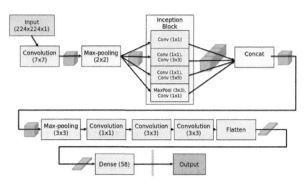

Fig. 2. Proposed CNN architecture

allows to find an optimal balance between the learning of the model and its evaluation on the test data. For our model, we started with a *batch_size* of 512 and a *learning_rate* of 10^{-3}, which was successively reduced to 256, 128 till 32 by decaying the learning rate to 10^{-5}. The choice of the hyper-parameters remains empirical, to the best of our knowledge, there is no exact approach automating the learning process. The hyper-parameters are selected according to the specificity of the network architecture and the type of data.

Fig. 3 shows the performance of the model on a balanced dataset of 116,000 images (2,000 images per class). We may observe that starting by the 10th epoch onwards, the loss function on evaluation (test) data goes to increase. In the field

TABLE II
INFERENCE ACCURACY ON THE DATA AND
VALIDATION SETS

	Dataset	Validation Data	Test Data
Nr of images	116,000	23,200	3,750
Top-1 Accuracy (%)	99.92	99.93	99.84

of machine learning, this case is called *overfitting*. Overfitting or overgeneralization is the situation in which the model learns well on the training data but does not generalize well on the test data. To avoid it, we early stop the training of the model and continue with the tune of hyperparameters, in our case, by setting the mini-batch size and learning rate smaller. In both graphs, we observe that from the 20th epoch onwards, the model learns well and the evaluation on the test data occurs correctly. Moreover, the descent of the loss functions decreases continuously until the end of the learning phase. We reach a predictive accuracy of our model of 99.9%.

Table II gives the performance of our model on our dataset and on validation data.

Now that we have a model that is highly accurate while relatively small in terms of parameters, we further limit its computation and storage needs by applying quantization.

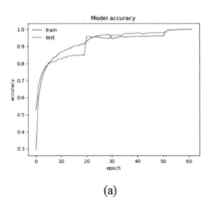

(a)

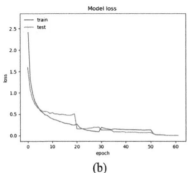

(b)

Fig. 3. Model scores: the evolution of (a) accuracy and (b) loss during the learning and testing processes

B. Quantification Principles

Quantization consists of reducing the number of bits necessary to represent a value. Its use in neural networks is not new [19], [20], but the introduction of deep convolutional neural network has however led to different works this past decade. There are now many different quantization approaches, ranging from quantizing only the parameters, quantizing both parameters (often only weights, not biases) and activations, quantizing on 16, 8, or even 2 or 1 bit, etc. The approaches using the smaller bit sizes are meaningful for hardware implementations only [21]–[24] to name a few. For the sake of this work, which targets of-the-shelf micro-controller-based boards, we will restrict ourselves to an 8-bit quantization of both the weights that is well suited to byte base computation in software, or with existing hardware accelerators (either ad-hoc or performing matrix-vector or matrix-matrix multiplications). As a result, the most demanding part of the neuron output computation. ($v_j = \sum_{i=0}^{n-1} x_i w_{ij}$) uses only 8-bit integer multiplications. This is key because the area and power complexity of a multiplier is in $O(b^2)$ where b is the number of bits of the inputs. Each multiplication produces a $2b$-bit result, that is accumulated with the adder to produce $(2b + \log_2 n)$-bit result, n being the number of inputs of the neuron. Using a 32-bit addition is a safe guess here, as there are very few chances that the accumulation takes place with more than 2^{16} inputs. It is also safe to have a bias b_j on 32-bit, as this is a single addition performed after all integer multiplications ($o_j = v_j + b_j$). It might even be the initial value of the accumulator.

As TensorFlow was the first framework to provide 8-bit integer arithmetic fine-tuned implementations for microcontrollers (using e.g., SIMD instructions) and Google TPU [25], we opted for using it given our high-power efficiency goal. We briefly summarize here the quantization approach that is advocated by and implemented in this framework, and thoroughly detailed in [26]. For a given convolutional layer, the quantization process produces in addition an offset (called zero-point, zp), and for each output channel of the layer a scale under the form of an integer multiplicand M and a shift s. The scale factor and offset must be applied before the activation function, leading (roughly, as the idea is to divide by 2^s which is not a raw shift for negative values) to $y_j = (o_j \times M \gg s) + zp$. These operations, done only once per kernel, are typically 32-bit, and the result is saturated to -128 or 127.

From a practical point of view, there are two main ways for quantizing a network: post-training quantization (PTQ) and quantization-aware training (QAT). PTQ consists of finding offsets and scale values to approximate the weights of an already trained network. Post-training works quite well on large networks, especially when lowering weight size to 8 bits or above. To further reduce bit size without incurring high accuracy losses, it is usually necessary to use QAT. This consists of training the network by considering the low precision behaviour during the process.

Google's TensorFlow-Lite (TF-Lite) open-source framework provides an API to convert and interpret quantized networks. Given our target that is microcontrollers possibly backed by an accelerator, for which lower than 8-bit precision is useless, we use the PTQ method. It produces weights and biases quantized to a fixed-point precision of 8-bit integer using the approach mentioned above and required by integer-only accelerators. PTQ takes a fully trained model and doesn't require additional modifications for conversion into a quantized model. Nevertheless, an important point for the conversion process is to provide a representative dataset, i.e., a small subset of the original dataset which covers the entire value space. This gives the quantization process the range of inputs values and it can then find the most appropriate 8-bit representation (multiplicand M and shift s) for each weight and activation value. To achieve the best possible performance, i.e., ensure that all computations are done using SIMD instructions or outsourced to the TPU, it is

TABLE III
INFERENCE ACCURACY ON THE DATA AND
VALIDATION SETS AFTER POST-TRAINING
QUANTIZATION

	Dataset	Validation Data	Test Data
Nr of images	116,000	23,200	3,750
Top-1 Accuracy (%)	97.58	97.23	96.18

recommended to strictly stick to the 8-bit data type. For this purpose, we perform full integer optimization with the TF-Lite converter, i.e., the inputs and the outputs use 8 bits.

The accuracy once the quantization process is given in Table III. There is a slight drop in accuracy, around 2% on the whole dataset, a bit less than 4% on the test data. The confusion matrix shows that errors fit into nearby classes, which is not perfect but reasonable from an applicative point of view. The experiments and results are presented next.

V. EXPERIMENTS

We implemented an end-to-end inference design based on our quantized neural network architecture using two boards representative of off-the-shelf IA edge platforms, one of them embedding a small TPU. These experiments aim at demonstrating that our solution is feasible on these types of low-power and limited resources devices. It shows in particular that both boards deliver performances that follow the pace of a production line test equipment with a good margin of progression. Indeed, test equipment usually produces batches of wafermaps, meaning that the performance of our solution target is a mean throughput of at least 1 inference per second, without a strong requirement on inference latency. Even though our solution is not intended to be integrated on a battery-powered embedded system,

power consumption still is an ongoing concern that must be considered. Power efficiency analysis shows as well that these performances are achieved with low-power consumption and good overall power efficiency.

These experiments are conducted using software implementation of our quantized neural network model. They are each using the available kernel implementation provided with their development kit without neither modification nor optimization from our side. Further optimization is surely possible, though we show through this type of experiment that optimizing only the neural network model is enough to deliver the required performances using general purpose hardware.

A. Experimental Setup

The two boards used in our experiments are Google's Coral standalone board and ST Microelectronics' STM32MP1 board. The Coral SoC embeds a quad Cortex-A53 and a Cortex-M4F as well as a small TPU coprocessor for neural network inference. The technical specification rates this TPU with 4 TOPS for 8-bit integer operations giving 2 TOPS per watt. The STM32MP1 board embeds a dual Cortex-A7 and a Cortex-M4.

Both boards are setup in the same way, as shown Figure 4. They are powered through a power-meter allowing to record power measures, in particular the total power consumption and curves of instantaneous power. For demonstration purposes, the whole setup is powered by a laptop on which an HDMI video capture card is connected to display the boards' display output. This capture device is not taken into account into the power measures.

Images being inferred are all already loaded into the boards' memory. This is consistent with the way wafermaps are generated in batches and not inferred

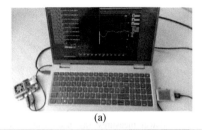

(a)

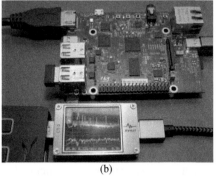

(b)

Fig. 4. Experiment setups:(a) Google's Coral setup (b) STM32MP1 setup. Both are powered through a small power-meter allowing to record the power consumption of the entire board

TABLE IV
CORAL BOARD: LATENCY MEASUREMENTS FOR A THROUGHPUT OF 10,000 IMAGES AND ON 5 CONSECUTIVE TESTS

Experiment	#1	#2	#3	#4	#5
Min Latency (ms)	0.959	0.956	0.956	0.952	0.952
Max Latency (ms)	2.916	3.568	4.160	4.278	3.825
Avg Latency (ms)	1.119	1.110	1.110	1.110	1.110

one by one along the day. Moreover, it would give results that would be hard to interpret as the pace and sizes of data batch transfers vary very much from one foundry to another.

B. Experimental Results

The inference latency is a measure to see the real time performance of the model. Experiments performed on 10,000 random images show that the average inference latency is about 1.11 ms. Table IV shows the inference latencies on 5 consecutive tests. Each time the maximum latency corresponds to the first inference which is longer due to the parameter caching of the model on the on-chip memory. Once the model has been loaded, the model weights are reused for the next inferences and so inference latencies are about three times lower.

Both boards achieve an accuracy of 97% confirming the numbers we measured on the host using the Tensor-

Flow framework. About inference time, the Coral reaches a rate of ≈903 inferences per seconds. The MP1, on its side performs, without hardware support for inference, at 5.5 inferences per seconds. These results are good regarding our target problem and are actually very satisfying considering the size of the data (224×224 pixels per image) and the number of classes (58). Real-time for wafer manufacturing means that we must keep the pace of the testing equipment within the production line, which yield a wafer every dozens of seconds at least. Making a minimum of around 5 inferences per seconds for the MP1 is thus well enough and gives a margin of progression both on production line speed and data complexity. For instance, it gives room for higher resolution wafermaps or more classes.

The power measures we have made show that the Coral board has an idling instantaneous power of 3.3 W while the MP1 stays at 1 W. This of course is due to the internal hardware: the Coral embeds a higher-grade processor and a hardware tensor processing unit. The Coral board has a cooling fan which might add to its overhead (the MP1 being a fan less board), although the fan never actually ran during our experiments. We though decided not to remove it from the equation to stay in a realistic use case, as it could be measure on an actual device attached to a machine in the production line. In other words, we measure the whole board components, including DRAM, peripheral accesses and the Linux kernel running on their cores, not specifically the coprocessor making the inference [27].

The left part of Figure 5 shows instantaneous power, both for idle and running state of each board. When performing an inference, the Coral board consumes a total of about 4.1 W while the MP1 runs at 1.3 W. It represents a 25% increase for the Coral and a 30% increase for the MP1. However, considering the error bars, both power increases can be considered similar. The right part of Figure 5 also shows an interesting indicator, the performance per watt, to compare how both boards perform and which one is the most efficient.

Table V gathers the numbers we get for each board. Results show without surprise that the Coral with its TPU is 164× faster than the MP1. When it comes to inferences per second per watt, the Coral performs better with 220, against 4.2 for the MP1. This gives a 52× better power-efficiency for the Coral board, which is easily explained by the dedicated ASIC for neural network acceleration. We still note that the TPU is not exploited at its maximum by the TensorFlow Lite backend, as the peak performance would lead to a 2 W increase in power consumption.

Similar or better efficiency might be achievable with low-power GPUs such as Nvidia Tegra. Though, we are not convinced that this is much interesting

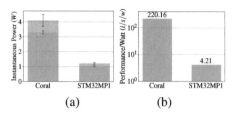

Fig. 5. Power measures: (a) Instantaneous power (idle on the bottom part of each bar) (b) Power efficiency in inference per second per watt

considering the key metric of total energy consumption, they would indeed run faster with an increased power of around 10W. However, the total consumed energy would be higher as we choose more and more power-hungry and fast hardware because of the nature of how batches of wafermaps are generated by test equipment. Faster devices would stay inactive most of the time, wasting their idle power waiting the next data batch. Thus, deliberately getting less throughput still delivers the required throughput for the industrial semi-conductor use case we consider, while being among the lowest power consumption solution we can get for this type of classification problem. A good future improvement of such solution would be to better tune hardware performances to either downclock, standby or even shutdown the inference platform at the right moment to save even more power. Finally, the market cost of such small boards such as the one used in these experiments are well under GPU solutions, making them even more attractive considering both the cost of the initial purchase and the maintenance replacement cost.

VI. CONCLUSION AND FUTURE WORK

Wafermap classification is an important step in semi-conductor process control. While the throughput of the test equipment is low compared to, e.g., video rate, it runs around the clock. Therefore,

TABLE V
PERFORMANCE AND POWER EFFICIENCY OF BOTH BOARDS

Board	Inference Performance	Average Power	Performance per Watt
Coral	902 i/s	4.1 W	220.16 i/s/W
STM32MP1	5.5 i/s	1.3 W	4.21 i/s/W

having a low footprint accurate power-efficient solution usable directly on the industrial machines is of interest to spare energy. To that end, we present in this paper a purpose defined neural network architecture that features a low parameter count, that we further quantize to limit the computation and memory resources necessary to perform inference. We implement this network on micro-controller boards with and without a hardware inference accelerator and show that it can perform inferences fast enough to follow test equipment pace, at a 4 inference per second per watt cost for a small microcontroller board and around 220 i/s/W using a low-cost embedded TPU accelerator.

We proposed an approach centered on neural network model optimization, demonstrating that it is a good approach toward low-power deep learning solutions. This approach can be applied to other industrial use-cases sharing the same dataset features. In particular, datasets with low interferences such as our black and white wafermaps generated by consistent test equipment are well suited. For instance, industries such as railway or photovoltaic manufacturing have test equipment generating very similar data with small changes between them. In the end, GoogLeNet, ResNet and others are very effective also with much more complex data such as 24-bit real life photographs, but they are overkill solutions when applied to very specific industrial applications. As promising as this seems, one of the most important aspects is the training dataset. A clean training dataset is an absolute necessity before applying any sort of deep learning approach and must be the first priority. This means that it must be large enough, well labeled and well balanced. Only that prerequisite enables efficient model optimization eventually allowing to downscale inference platforms.

With an appropriate model, further optimization can be made by focusing on the actual inference implementation. First, kernel implementation can be optimized with power usage in mind. For instance, some instructions are by nature more power consuming than others, such as memory load and stores, or branching instructions. Using works focusing on instruction-level power consumption optimization could thus be used to trade more power efficiency against performance, e.g., by redoing computation rather than storing and then loading an intermediate result. Secondly, hardware implementation solutions allow to optimise inference power efficiency even further. Multiple level of abstraction can be used depending on how much work we are willing to put in such implementations. It extends from High Level Synthesis (HLS) solutions to HDL solutions. The interesting point with hardware solutions is that model quantization can be further pushed toward ternary or binary models, as demonstrated in other application domains [28]. This allows very efficient matrix multiplications saving even more power while accuracy is only slightly degraded.

ACKNOWLEDGEMENTS

The authors would like to thank Maxime Martin, STMicroelectronics, Crolles, for providing them the dataset and process related information.

F. Pétrot would also like to acknowledge the support of the French Agence Nationale de la Recherche (ANR) through the MIAI@Grenoble Alpes ANR-19-P3IA-0003 grant.

References

[1] B. T. Murphy, "Cost-size optima of monolithic integrated circuits," Proceedings of the IEEE, vol. 52, no. 12, pp. 1537-1545, 1964.

[2] K. Park and H. Simka, "Advanced interconnect challenges beyond 5nm and possible solutions," in 2021 IEEE International Interconnect Technology Conference (IITC). IEEE, pp. 1-3, 2021

[3] M. H. Hansen, V. N. Nair, and D. J. Friedman, "Monitoring wafer map data from integrated circuit fabrication processes for spatially clustered defects," Technometrics, vol. 39, no. 3, pp. 241-253, 1997.

[4] F. Duvivier, "Automatic detection of spatial signature on wafermaps in a high volume production," in International Symposium on Defect and Fault Tolerance in VLSI Systems. IEEE, pp. 61-66, 1999.

[5] A. Krizhevsky, I. Sutskever, and G. E. Hinton, "Imagenet classification with deep convolutional neural networks," Advances in Neural Information Processing Systems, vol. 25, 2012.

[6] M.-J. Wu, J.-S. R. Jang, and J.-L. Chen, "Wafer map failure pattern recognition and similarity ranking for large-scale data sets," IEEE Transactions on Semiconductor Manufacturing, vol. 28, no. 1, pp. 1-12, 2015.

[7] T. Nakazawa and D. V. Kulkarni, "Wafer Map Defect Pattern Classification and Image Retrieval Using Convolutional Neural Network," IEEE Transactions on Semiconductor Manufacturing, vol. 31, no. 2, pp. 309- 314, 2018.

[8] M. Saqlain, Q. Abbas, and J. Y. Lee, "A Deep Convolutional Neural Network for Wafer Defect Identification on an Imbalanced Dataset in Semiconductor Manufacturing Processes," IEEE Transactions on Semiconductor Manufacturing, vol. 33, no. 3, pp. 436-444, 2020.

[9] J. Shim, S. Kang, and S. Cho, "Active Learning of Convolutional Neural Network for Cost-Effective Wafer Map Pattern Classification," IEEE Transactions on Semiconductor Manufacturing, vol. 33, no. 2, pp. 258- 266, 2020.

[10] R. Wang and N. Chen, "Defect Pattern Recognition on Wafers using Convolutional Neural Networks," Quality and Reliability Engineering International, vol. 36, no. 4, pp. 1245-1257, 2020.

[11] T.-H. Tsai and Y.-C. Lee, "A light-weight neural network for wafer map classification based on data augmentation," IEEE Transactions on Semiconductor Manufacturing, vol. 33, no. 4, pp. 663-672, 2020.

[12] H. Kahng and S. B. Kim, "Self-supervised representation learning for wafer bin map defect pattern classification," IEEE Transactions on Semiconductor Manufacturing, vol. 34, no. 1, pp. 74-86, 2021.

[13] L. Andrade, T. Baumela, F. Pétrot, D. Briand, O. Bichler, and M. Coppola, Efficient Deep Learning Approach for Fault Detection in the Semiconductor Industry, ser. Series in Communications and Networking. River Publishers, ch. 2.2, pp. 131-146, 2021.

[14] C. Cortes, L. D. Jackel, and W.-P. Chiang, "Limits on learning machine accuracy imposed by data quality," Advances in Neural Information Processing Systems, vol. 7, 1994.

[15] D. H. McLain, "Two dimensional interpolation from random data," The Computer Journal, vol. 19, no. 2, pp. 178-181, 1976.

[16] F. Hutter, L. Kotthoff, and J. Vanschoren, Automated machine learn-

ing: methods, systems, challenges. Springer Nature, 2019.

[17] T. Elsken, J. H. Metzen, and F. Hutter, "Neural architecture search: A survey," The Journal of Machine Learning Research, vol. 20, no. 1, pp. 1997-2017, 2019.

[18] C. Szegedy, W. Liu, Y. Jia, P. Sermanet, S. Reed, D. Anguelov, D. Erhan, V. Vanhoucke, and A. Rabinovich, "Going deeper with convolutions," in Proceedings of the IEEE conference on computer vision and pattern recognition, pp. 1-9, 2015.

[19] G. Dundar and K. Rose, "The effects of quantization on multilayer neural networks," IEEE Transactions on Neural Networks, vol. 6, no. 6, pp. 1446-1451, 1995.

[20] B. Hoskins, M. Haskard, and G. Curkowicz, "A vlsi implementation of multi-layer neural network with ternary activation functions and limited integer weights," in 1995 20th International Conference on Microelectronics, pp. 843-846, 1995.

[21] R. Andri, L. Cavigelli, D. Rossi, and L. Benini, "Yodann: An ultra-low power convolutional neural network accelerator based on binary weights," in IEEE Computer Society Annual Symposium on VLSI, pp. 236-241, 2016.

[22] Y. Umuroglu, N. J. Fraser, G. Gambardella, M. Blott, P. Leong, M. Jahre, and K. Vissers, "Finn: A framework for fast, scalable binarized neural network inference," in Proceedings of the 2017 ACM/SIGDA International Symposium on Field-Programmable Gate Arrays, pp. 65-74, Feb. 2017.

[23] A. Prost-Boucle, A. Bourge, and F. Pétrot, "High-efficiency convolutional ternary neural networks with custom adder trees and weight compression," ACM Transactions on Reconfigurable Technology and Systems, vol. 11, no. 3, pp. 1-24, 2018.

[24] R. Zhao, W. Song, W. Zhang, T. Xing, J.-H. Lin, M. Srivastava, Gupta, and Z. Zhang, "Accelerating binarized convolutional neural networks with software-programmable fpgas," in Proceedings of the 2017 ACM/SIGDA International Symposium on Field-Programmable Gate Arrays. ACM, pp. 15-24, 2017.

[25] N. P. Jouppi, D. H. Yoon, M. Ashcraft, M. Gottscho, T. B. Jablin, G. Kurian, J. Laudon, S. Li, P. Ma, X. Ma et al., "Ten lessons from three generations shaped google's tpuv4i: Industrial product," in 2021 ACM/IEEE 48th Annual International Symposium on Computer Architecture (ISCA). IEEE, pp. 1-14, 2021.

[26] B. Jacob, S. Kligys, B. Chen, M. Zhu, M. Tang, A. Howard, H. Adam, and D. Kalenichenko, "Quantization and training of neural networks for efficient integer-arithmetic-only inference," in IEEE conference on computer vision and pattern recognition, pp. 2704-2713, 2018.

[27] V. Sze, Y.-H. Chen, T.-J. Yang, and J. S. Emer, "How to evaluate deep neural network processors: Tops/w (alone) considered harmful," IEEE Solid-State Circuits Magazine, vol. 12, no. 3, pp. 28-41, 2020.

[28] A. De Vita, D. Pau, L. Di Benedetto, A. Rubino, F. Pétrot, and G.D. Licciardo, "Low power tiny binary neural network with improved accuracy in human recognition systems," in 23rd Euromicro Conference on Digital System Design. IEEE, pp. 309-315, 2020.

Ana Pinzari received the Ph.D degree in computer science at Université de Technologie de Compiègne, Compiègne, France in 2012, on the subject of defining a methodology for the execution of real-time image processing algorithms. Then, she worked in a scientific research organization ECSI (European Electronic Chips & Systems Initiative) focusing on new design methods, tools and standards for design of complex electronic systems. Besides dissemination activities of European EDA research projects, she took part in definition and implementation of renowned international conferences focusing on specification and design languages (FDL, DVCon Europe), signal and image processing (DASIP) and debug (S4D). She joined Grenoble INP, TIMA laboratory in 2021, as a post-doctoral research engineer. Her current research interest is in exploring optimization methods for the hardware implementation of neural networks.

Thomas Baumela received his Ph.D degree in computer science at Université Grenoble Alpes, Grenoble, France in 2021, working on the hardware and software integration of devices in FPGA using a co-designed message-based approach at the TIMA Laboratory. He then stayed at the TIMA laboratory as a post-doctoral researcher, working on implementation and integration of hardware accelerated neural networks.

Liliana Andrade received her engineering degree from Universidad de Los Andes, Mérida, Venezuela, in 2012; and her Ph.D. degree in computer science, telecommunications and electronics from Université Pierre et Marie Curie (Paris VI), Paris, France, in January 2016. Since September 2017, she is with the TIMA laboratory, System Level Synthesis team, in Grenoble, France. She is associate professor in computer science at the Polytechnical School of Université Grenoble Alpes. Her research interests include system-level modeling, design and validation of systems

on chip, and acceleration of heterogeneous systems simulation.

Marcello Coppola is technical Director at STMicroelectronics. He has more than 25 years of industry experience with an extended network within the research community and major funding agencies with the primary focus on the development of break-through technologies. He is a technology innovator, with the ability to accurately predict technology trends. He is involved in many European research projects targeting Industrial IoT and IoT, cyber physical systems, Smart Agriculture, AI, Low power, Security, 5G, and design technologies for Multicore and Many-core System-on-Chip, with particular emphasis to architecture and network-on-chip. He has published more than 50 scientific publications, holds over 26 issued patents. He authored chapters in 12 edited print books, and he is one of the main authors of "Design of Cost-Efficient Interconnect Processing Units:Spidergon STNoC" book. Until 2018, he was part of IEEE Computing Now Journal Technical editorial board. He contributed to the security chapter of the Strategic Research Agenda (SRA)to set the scene on R&D on Embedded Intelligent Systems in Europe. He is serving under different roles numerous top international conferences and workshops. Graduated in Computer Science from the University of Pisa, Italy in 1992.

Frédéric Pétrot received the Ph.D. degree in computer science from Université Pierre et Marie Curie (now Sorbonne Université), Paris, France, in 1994, working on the Alliance CAD system for VLSI design. He became assistant professor there in 1995, and worked on higher level of abstractions for ESL. He joined Grenoble-INP/Ensimag, Grenoble (now Institute of Engineering Univ. Grenoble Alpes), France, as a professor in 2004. His research takes place in the TIMA Laboratory and focuses on the specification, simulation, and implementation of multiprocessor systems on chip architectures, including circuits, system software, and Computer-Aided-Design aspects. Since 2019, he holds the Digital HW AI Architectures chair of Grenoble Multidisciplinary Institute in Artificial Intelligence.

Low-power Analog In-memory Computing Neuromorphic Circuits

Roland Müller, Bijoy Kundu, Elmar Herzer, Claudia Schuhmann, and
Loreto Mateu

Abstract—The presented neuromorphic circuits comprise synaptic weights and neurons including batch normalization, activation function, and offset cancelation circuits. These neuromorphic circuits comprise an effective 3.5 bits weight storage based on binary memory cells while the analog multiplication and addition operation is based on a voltage divider principle. To experimentally proof the working principle, three fully connected layers (50x20, 20x10 and 10x4) have been designed. The connection between these layers is realized completely in the analog domain without ADCs and DACs in between. An inference state machine takes care of pipelining the layers for a proper operation during inference. The schematic and layout of the neuromorphic circuits comprised in these layers have been automatically generated with a in-house designed automation framework. This framework, called UnilibPlus, is a Python-based Cadence Virtuoso add-on. Simulation results of weight loading, transfer of input values, inference and read inference results via SPI interface show a correct operation of the designed ASIC with 12 nJ per inference and 5 μs latency.

The research leading to these results has received funding from the Electronic Components and Systems for European Leadership Joint Undertaking under grant agreement No 826655 – project TEMPO. This Joint Undertaking receives support from the European Union's Horizon 2020 research and innovation programme and Belgium, France, Germany, Netherlands, Switzerland. TEMPO has also received funding from the German Federal Ministry of Education and Research (BMBF) under Grant No. 16ESE0405. The authors are responsible for the content of this publication. Roland Müller, Bijoy Kundu, Elmar Herzer, Claudia Schuhmann and Loreto Mateu are with the Advanced Analog Circuits group at Fraunhofer IIS, Am Wolfsmantel 3, 91058 Erlangen.

Index Terms—AI accelerators, Application specific integrated circuits, Analog processing circuits, Neural networks, Neural network hardware, SRAM cells

I. INTRODUCTION

THE outperforming results of deep neural networks (DNNs) for cloud based platforms has pushed the development of new DNN models and hardware components for edge devices. Since the requirements for cloud and edge computing in terms of performance are different, new architectures are explored. Inference accelerators for edge applications target low energy consumption results per inference. Since the multiply-accumulate (MAC) operation together with the activation function are the main operation to perform in DNNs, inference accelerators perform efficiently these operations. This work explores a fully analog implementation approach for DNNs by using SRAM cells for the storage of the synaptic weights in the crossbar arrays in 28-nm GlobalFoundries technology. This work presents: (i) a crossbar implementation per neural network layer with 3.5 bit accuracy for weight storage (equivalent to three positive, three negative and zero weight values); (ii) a pipeline between layers arranged via finite state machines (FSMs); (iii) the use of SRAM cells for in-memory computing which ensures the porting of the circuit to other technology nodes and reduces data movement; (iv) a voltage divider approach for the synaptic weight circuit; (v) a fully analog design of the neural network without ADC and DACs between layers in order to get rid of their overhead [1] for the DNN design; (vi) a weight loading via SPI interface to update the synaptic weights of the layers; and (vii) circuit blocks included to enable the test of all neural network building blocks.

The paper is organized as follows: Section II provides an overview of the ASIC architecture, a short description at block level as well as a more detailed explanation of the synaptic and neuron circuits. Section III summarizes the simulation results obtained at neuron level and during inference. Design for testing techniques have been used in the ASIC design to be able to scan the correct functionality of each one of the circuits that conform the crossbar arrays of the three layers separately. Section IV provides an overview of the functional tests and evaluation tests strategy for the verification of the designed circuits and evaluation of their performance. Finally, Section V provides a summary of the remaining work and concluding remarks.

II. ARCHITECTURE FOR AN ANALOG IMPLEMENTATION OF NNs

A. Overall Architecture

Fig. 1 shows the top-level block diagram of the inference accelerator ASIC, which consists of a *Digital Control* block, a *Frontend* circuit, a 3-layer *Deep Neural Network (DNN)*, a *Test Multiplexer* and a *Capture Stage*. The *Digital Control* block includes a Serial Peripheral Interface *(SPI)* to control the configuration of the ASIC and to enable the transfer of input and output data during inference. Additionally, three different types of finite state machines (FSMs) are included: a *Frontend-FSM (FE-FSM)*, three

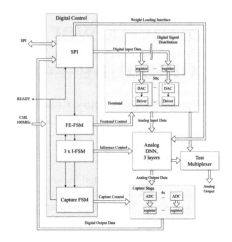

Fig. 1. Top-level block diagram of the DNN circuit including digital control and circuitry used to test the ASIC.

Inference-FSMs (3x I-FSM) and a *Capture FSM*. These state machines control the execution and timing of the inference executed in the DNN. The *Frontend* block distributes the 8-bit serial input data to registers for each input of the first DNN layer. The registers are connected to a combination of a *DAC* and a *Driver* circuit. The *DAC* converts the digital input data to the analog domain whilst the *Driver* provides a low output resistance to drive the input resistance of the DNN. The outputs of the *Frontend* are then used as the *Analog Input Data* to the DNN with the DNN containing three fully connected (FC) layers with input-output dimensions of 50×20, 20×10 and 10x4. After all layers have finished processing its inputs, the *Analog Output Data* of the DNN is then captured by four 8-bit ADCs (one per output neuron of the last layer) implementing the conversion back from analog to digital domain. These *Digital Output Data* can then be read out via SPI. Finally, the *Test Multiplexer* provides access to all frontend and neuron output voltages for measurement and verification purposes.

The architecture of the chip enables the measurement, evaluation, and verification of almost every building block within the DNN separately. Therefore, block level simulation results can be correlated with the respective measurements. The results can further be used to extrapolate the circuit's behavior to bigger scale networks that shall be implemented.

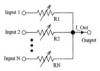

Fig. 2. Synaptic weight circuit using variable resistors.

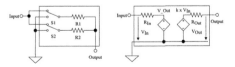

Fig. 3. Voltage divider synaptic weight implementation: (a) schematic and (b) equivalent circuit illustration.

B. Voltage Divider Approach For Synaptic Weights

Synaptic weights in analog DNN accelerators are one of the major components in such circuits. They can be implemented in multiple ways with the standard being a voltage to current conversion using a variable resistor like shown in Fig. 2 [1]. In this implementation, the voltages across the resistors $R1$ to RN are converted into a current by the variable resistors, with the resistor values being set to a certain value provided by the neural network model. The resulting currents are then summed up to the final output current. This implementation requires the voltage at the *Output* node to be constant (virtual ground) and the current to be converted back to a voltage (I-V conversion) to forward the calculated value as *Input* voltage to the next layer. Different approaches can be followed to implement the virtual ground and the I-V conversion. For example, a transimpedance amplifier (TIA) or a shunt resistor can be used. The latter one can only approximately provide a virtual ground if its resistance is small enough which would lead to very low output voltages. Using a TIA, the virtual ground requirement is fulfilled but it must feedback the same current as *I_Out*. This leads to increased energy consumption whilst stability proves to be critical as the resistance connected to its input is dependent on the used weight values.

Because of the previously mentioned issues, a different approach has been implemented in our ASIC design. For this, we use a configurable voltage divider [2] as shown in Fig. 3 replacing the variable resistor circuit.

Each synaptic weight circuit consists of two resistors $R1$ and $R2$. Both resistors can be connected either to the *Input* node or to the analog ground. This is illustrated in Fig. 3(a) by the switches $S1$ and $S2$. The resistors are sized such that the output resistance R_{Out}, illustrated in Fig. 3(b) stays constant regardless of the switch configuration whilst the weight value k is equivalent to voltage divider ratio. Table I shows the possible weight values k and the corresponding input resistance R_{In} for the four different switch configurations under the condition that $R1 = 2 \cdot R2$.

TABLE I
WEIGHT VALUE AND INPUT RESISTANCE OF THE
VOLTAGE DIVIDER SYNAPTIC WEIGHT

S1	S2	k	R_{In}
Ground	Ground	0	-
Input	Ground	0.33	R1
Ground	Input	0.67	R2
Input	Input	1	R1 ∥ R2

$$V_{OUT} = \frac{1}{N} \sum_{i=1}^{N} V_{In,i} \cdot k_i, \qquad (1)$$

where N is the number of inputs connected together to a common output through synaptic weights, $V_{In,i}$ is the input voltage for *Input i*, k_i are the weight values for *Input i*, as shown in Table I.

If multiple such synaptic weight circuits are connected in parallel to form a crossbar array output, the output voltage is defined by (1). The output voltage is the average of the weighted sum of the input voltages, which leads to a reduction of the effective weight values. Therefore, a recovery gain needs to be applied to the outputs so as to bring them to a usable range.

The voltage divider approach for the synaptic weight does not require an I-V conversion because every output of the crossbar array is a voltage. This improves the accuracy compared with the current based solution, because the weight value only depends on the matching of unit resistors. The energy consumption is also reduced because only a voltage amplifier is needed which can be implemented in low power switch capacitor topology.

Further, as the output is a voltage, the signal routing within the crossbar is simplified as parasitic resistances at the outputs do not influence the calculation result.

Finally, to implement both positive and negative weight values, the synaptic weight circuit was replicated to create a differential output pair like shown in Fig. 4. If a positive weight value is required, switches *S1* and *S2* are used to set the weight values, whilst switches *S3* and *S4* connect the resistors to *R3* and *R4* to the analog ground voltage. For negative weight values, the switches *S3* and *S4* set the required weight values whilst the *R1*

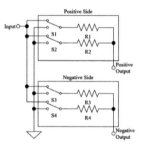

Fig. 4. Differential voltage divider synaptic weight for positive and negative weight values.

and *R2* are connected to analog ground via *S1* and *S2*.

C. Neuron Circuit

As stated in section B, the neuron circuit must implement a gain due to the averaging operation in the crossbar. Additionally, a batch normalization circuit must be implemented which consists of a variable gain with a following offset addition. Finally, a defined nonlinearity is required to perform the activation function. Equation (2) defines the operation that the neuron implements

$$V_{Neuron,Out} = f(V_{Neuron,\ In} \cdot k_{Neuron} + V_{Offset}) \qquad (2)$$

where $V_{Neuron,In}$, is the input voltage, $V_{Neuron,Out}$ is the output voltage, V_{Offset} is the batch normalizations offset voltage and k_{Neuron} represents the combination of the mentioned recovery gain and the batch normalizations variable gain. The function f is defined as the nonlinear activation function, which in this case is a Rectified Linear Unit (ReLU) function.

The neuron circuit comprises two amplification stages: a switched capacitor variable gain stage [4] to implement the batch normalization circuit and a buffer stage including the ReLU function. In between both stages, a sample and hold

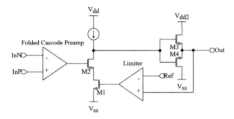

Fig. 5. ReLU activation function and buffer circuit.

stage is placed to decouple and pipeline the operation of the layers in the neural network.

As stated, the activation function should be a ReLU function, which has a constant output value for negative input values and a linear rising output value for positive input values. Ideally, the maximum output value is not limited but the output voltage cannot exceed its positive supply voltage. Therefore, for high input voltages, the output voltage will result in a constant value. This behavior equals the DC transfer function of an operational amplifier, which was therefore chosen as the architecture for the activation function. The output range reaches from 200 mV to 400 mV at a supply voltage of 1V. The limits were chosen such that there is enough headroom to the negative supply voltage and that it does not exceed the half of the supply voltage. This eases the requirement for the switch resistance in the synapse circuits since an NMOS switch is sufficient.

Fig. 5 shows a simplified schematic of the implemented ReLU activation function. Transistors $M3$ and $M4$ form a source follower output stage. This configuration was selected due to the relatively low load resistance from the following layer of approximately 1 kΩ. The upper limit is defined by V_{dd2} which is set to 400 mV and is therefore limited by the supply voltage. The lower limit cannot be generated in the same manner because

the input voltage of the output stage is limited to V_{ss}. Therefore, the gate-source voltage of transistor $M4$ would be 0 if the required output voltage is also 0. In this case, the output voltage cannot reach 0 if a load resistance is connected, as transistor $M4$ would not drive any current. To avoid this effect, a limiting amplifier (*Limiter*) was introduced that compares the output voltage to a reference voltage (*Ref*) of 200 mV and sets the output voltage to the reference voltage in case the output voltage drops below the reference voltage [5].

The transition between the linear and the limited region of the activation function is implemented by the two-gate amplification stage, which consists the transistors $M1$ and $M2$. In the linear region, $M1$ is turned completely on, leaving the *Limiter* with no effect on the circuit behavior. In the region where the *Limiter* is active, the current through $M1$ is regulated such that the output voltage stays constant.

III. SIMULATION RESULTS

Corner and mismatch simulations have been performed for all analog circuits. Digital simulations for the digital core including the SPI module and FSMs have also been performed. This section provides further details of the simulations results obtained for the neuron circuit and for the inference of the 3 layers.

A. Neuron

The neuron circuit was simulated for different operating conditions including different batch normalization gain and offset values, load resistances, load voltages and input voltages. Fig. 6 shows the simulation results for the transfer function of the neuron for different process, voltage, and temperature (PVT) variations, with 5% supply voltage variation

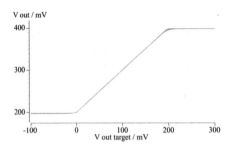

Fig. 6. Transfer function of the neuron over PVT variations.

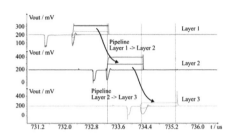

Fig. 7. Transient simulation results of the output voltages of all three layers.

and a temperature range of 0°C to 70°C. As required, the output voltage stays constant for negative input voltages, rises linearly for positive voltages and is clipped for differential input voltages exceeding 200 mV. Further, the transitions between linear and constant regions are well defined which is also required for a successful training and inference of a neural network.

B. Inference

The inference was simulated with multiple random weight and input value patterns. In a first step, the weights are loaded onto the neural network and afterwards, multiple inferences have been executed with different input data value sets. Fig. 7 shows the transient simulation results for a single inference, where each plot shows the output voltages of the neurons per layer. The operation of each neuron starts with an offset sampling phase. During this phase, the calculations in the batch normalization circuits are executed, and its result is sampled on an internal capacitor. Afterwards, the activation function is calculated leading to a valid output value to be used in the next layer for calculating the multiply-and-accumulate results. The same process is executed in each layer of the neural network in a pipelining process described in Section II.C. Within the neuron, the input

voltage is amplified and then sampled in a sample-and-hold stage. Right after the sampling happened in one layer, the preceding neurons are switched off to reduce the energy consumption.

The three vertical markers in Fig. 7 highlight the time points at which the result of each layer is valid. The neuron output voltages at these points have been compared to an ideal calculation executed in Python. The average difference resulted in 1.9 mV and a maximum difference in 6.9 mV between ideal calculations and simulations. As such, we derive 1% average and 3% maximum difference for a neuron output range of 200 mV.

C. Key Performance Indicators

Table II summarizes the main performance indicators of the work presented in the paper. The energy consumption of the neural network was estimated based on a Python script containing electrical simulation data of the neuron energy consumption in relation to its operation conditions. Then, the operating conditions for each neuron in the neural network are calculated for different random weight and input pattern. These operating conditions are then used to get the respective energy consumption for each neuron from the previous characterization. Finally, all the energy consumption values are summed which leads to the

TABLE II
KEY PERFORMANCE INDICATORS FOR THIS
WORK

Parameter	Value
CMOS Technology	28-nm GF
Synapse Memory Size	4.96 Kb
Weight precision	3.5b (7 levels)
Core Area	940 μm x 890 μm
Power Supply	1.0 V
Energy/Inference	12 nJ
Latency	5 μs
Power Efficiency	193 GOPS/W

final energy consumption of the complete neural network with an average result of 12 nJ.

IV. EVALUATION AND VERIFICATION

The inference accelerator ASIC is designed as a test chip and provides access to all neuron outputs via the test multiplexer as illustrated in Fig. 1. To verify the functionality and evaluate the performance of all circuit components based on measurements, test patterns are required. These test patterns are automatically generated by a Python script that creates the required files for weight loading and inference. In addition, the expected neuron output voltages are calculated for later comparison and validation. This section describes the tests and test patterns used for evaluation and verification of the ASIC.

A. Functional Test

Functional tests validate the expected operation of all circuit building blocks. Therefore, the test patterns must configure the ASIC in such a way, that one component is tested at a time.

Frontend Test

First, the digital input value for one of the frontend DACs at time is swept through its complete range whilst the frontend drivers output voltage is measured and compared with the expected value.

Neuron Test

In a second test, each neuron is validated individually. This is done by setting all the weight values to 0 which ensures that there is no influence between the neurons. Afterwards, the batch normalization offset is swept through its complete range, which leads to a varying output voltage of the neuron. For each neuron and each batch normalization offset value, the neuron output voltage is measured and compared with the expected value.

Weight Test

After testing the neuron circuit´s functionality, the weights can be tested. This is done by setting the output voltage of a frontend driver or a neuron in a layer to the center of its output range value whilst sweeping one of the connected weights through its possible values and measuring the output voltage of the following neuron.

In case of negative weight values and a weight value of 0, the neuron´s output voltage would always be 200 mV due to the limiter circuit. Therefore, the batch normalization offset value is used in this case to set the neuron´s output voltage to a value greater than 200 mV. The different combinations used are summarized in Fig. 8.

Batch Normalization Gain Test

The final functional test verifies the batch normalization gain configuration by sweeping the gain value through its overall range. In this test, according to the selected gain value, one or more synaptic weights must be used to set the

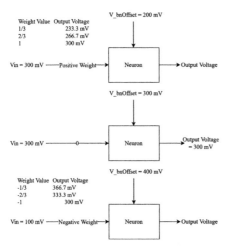

Fig. 8. Different combinations of weight value, batch normalization offset and resulting output voltage to test the synaptic weights.

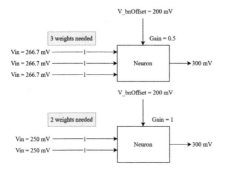

Fig. 9. Different number of weights in use for different gain configurations.

neurons output voltage to a measurable value. The number of weights is selected such that the output voltage of the currently tested neuron is close to 300 mV and the input voltage to the weights in use never exceeds 300 mV. Fig. 9 shows two different configurations for a gain value of 0.5 and 1 where three and two synaptic weights are used.

B. Evaluation Test

The evaluation tests first verify if the complete inference is executed correctly.

Secondly, the energy consumption in different configurations is measured.

Random Test Pattern

The first evaluation test uses random weight and input value patterns. For each weight pattern, multiple input value patterns are used. This test executes inferences with pseudo realistic values and therefore is a good indicator for the energy consumption in realistic scenarios. Furthermore, its results can be used to verify the calculated energy consumption results described in section III.B.

Best and Worst Case Energy Consumption

The final test configures the ASIC in such a way that the lowest and highest energy consumption is achieved. On the one hand, the test case with lowest energy consumption occurs when all weights are set to 0 whilst the digital input values are also 0. On the other hand, the highest energy consumption requires half of the weights in each column of the crossbar array to be set to their maximum value whilst the other weights are set to their minimum value with the neural networks input voltages and the batch normalization offset values all set to their maximum value.

V. CONCLUSIONS AND FUTURE WORK

While this work shows the architecture and circuits implemented for a fully analog in-memory computing implementation of a DNN based on SRAM cells, the simulation results as well as the test and evaluation strategy, measurements are still pending and will be executed as soon as the ASIC is fabricated and packaged.

The SRAM-based in-memory computing analog DNN implementation presented in this work shows good energy efficiency performance, which makes it attractive for small models with fixed sizes of neurons per layer. In comparison with many emerging non-volatile memory (eNVM) technologies like FeRAM, PCM, RRAM or FeFETs, SRAM cells are a mature technology that can be implemented in any technology node. While those eNVMs does not add any leakage to the overall ASIC while no inference is performed, SRAM is a good choice for applications in which inferences are continuously running.

The circuits presented here have been simulated across corners and mismatch and provide a stunning accuracy while comparing their results with ideal DNN models. Thus, the synaptic divider approach and overall architecture proves that a fully analog computation of neural networks can be implemented.

REFERENCES

[1] A. Shafiee, A. Nag, N. Muralimanohar, R. Balasubramanian, J. P. Strachan, M. Hu, R. S. Williams, and V. Srikumar. "ISAAC: a convolutional neural network accelerator with in-situ analog arithmetic in crossbars", SIGARCH Comput. Archit. News 44, no. 3, pp. 14–26, June 2016, https://doi.org/10.1145/3007787.3001139.

[2] E. Herzer and R. Müller, "Electronic Circuit for Calculating Weighted Sums," EP3992862, May., 4, 2022.

[3] R. Müller et al., "Hardware/Software Co-Design of an Automatically Generated Analog NN", in Proc. SAMOS 2021, Lecture Notes in Computer Science, vol. 13227, pp. 385–400, Springer, Cham, https://doi.org/10.1007/978-3-031-04580-6_26

[4] C. Enz and G. C. Temes, "Circuit techniques for reducing the effects of op-amp imperfections: autozeroing, correlated double sampling, and chopper stabilization," in Proc. of the IEEE, vol. 84, no. 11, pp. 1584-1614, Nov. 1996, doi: 10.1109/5.542410.

[5] C. Cini, C. Diazzi and P. Erratico, "Limited output operational amplifier", U.S. Patent vol. 4 pp. 672–326, June, 9, 1987.

Roland Müller born on the 19. January 1994 obtained his bis B.Eng. at the OTH Regensburg in 2017 and his M.Sc. in 2019 at the FAU Erlangen, both in electrical engineering.

In May 2019, he joined the department of Integrated Circuits and Systems at Fraunhofer IIS, Erlangen (Germany), where he is working in the field of analog-mixed signal design of neural network accelerators and design automation for such circuits. Currently, he is pursuing his PhD. His main research interests include low power analog-mixed signal circuits, neuromorphic computing and electronic design automation.

Bijoy Kundu received B.Eng. degree in Electronics and Telecommunication Eng. from IIEST, Shibpur, India, in 2013, and M.Sc. degree in Information and Communication Eng. from TU Darmstadt, Germany, in 2017.

He joined Fraunhofer IIS as a research engineer in 2017 and is currently working there in the Advanced Analog Circuits group. His main research interests include low power analog-mixed signal circuits, data converters, and neuromorphic computing.

Elmar Herzer received his Dipl.-Ing. degree in electrical engineering from the University of Stuttgart, Germany, in 1997.

Since 1997 he has been with the department for analog and mixed signal IC Design, at Fraunhofer IIS, Erlangen. He has been involved in numerous ASIC projects with mixed signal sensor frontends for automotive, industrial and consumer applications, mainly with 3D-Hall sensors, where he focused on low noise pre-amplification and Delta-Sigma ADCs. As a chief engineer he guided the 22 nm SOI design enablement and IIS 22 nm design flow. He is working currently in the field of analog-mixed signal design of neural network accelerators as a system architect. His main research interests include low noise analog-mixed signal circuits, approximate computing and analog design automation. He has filed 12 patents.

Loreto Mateu obtained her B.S. in Industrial Engineering in 1999, her M.S. in Electronic Engineering in 2002 and her PhD degree in June 2009 with a thesis titled Energy Harvesting from Human Passive Power at the Universitat Politècnica de Catalunya.

In June 2007, she joined Fraunhofer IIS as research engineer and became chief scientist in 2012. Since 2018 she is group manager of the Advanced Analog Circuits group. Her research interests include ultra-low power design, AC-DC and DC-DC converters, neuromorphic hardware and energy harvesting. She holds several patents and is co-author of a book, a book chapter and international papers published journals and conference proceedings.

Claudia Schuhmann obtained her diploma in physics in 1986 at the FAU Erlangen.

Until 1995 she worked on analogue design implementation in the IC department of SIEMENS AG. In 1999 she joined Fraunhofer IIS where she is working in the field of digital design. Her main focus is on the implementation and physical verification of digital and mixed signal designs in deep submicron technologies.

Tools and Methodologies for Edge-AI Mixed-Signal Inference Accelerators

Loreto Mateu, Johannes Leugering, Roland Müller, Yogesh Patil, Maen Mallah, Marco Breiling, and Ferdinand Pscheidl

Abstract—The ANDANTE project aims to tackle the hardware/software co-design challenge that arises from the development of novel (neuromorphic) edge-AI accelerators. For this purpose, Fraunhofer IIS and EMFT among other partners developed several tools to facilitate design, training and deployment of artificial neural networks in dedicated hardware accelerators. These tools provide hardware-aware training, automatic hardware generation, compilers, estimation of KPIs like energy consumption, and simulation under consideration of the constraints imposed by the targeted hardware implementation and use cases. The development of such a tool chain is a multidisciplinary effort combining neural network algorithm design, software development and integrated circuit design. We show how such a toolchain allows to optimize and verify the hardware design, reach the targeted KPIs, and reduce the time-to-market.

Index Terms—AI accelerators, analog processing circuits, application specific integrated circuits, mixed-signal simulation, neural networks, neural network hardware, neuromorphic computing, system on chip.

The research leading to these results has received funding from the Electronic Components and Systems for European Leadership Joint Undertaking under grant agreement No 876925 – project ANDANTE. This Joint Undertaking receives support from the European Union's Horizon 2020 research and innovation programme and France, Belgium, Germany, The Netherlands, Portugal, Spain, Switzerland. ANDANTE has also received funding fro the German Federal Ministry of Education and Research (BMBF) under Grant No. 16MEE0116 and 16MEE0117. The authors are responsible for the content of this publication. Loreto Mateu, Johannes Leugering, Roland Müller, Yogesh Patil, Maen Mallah and Marco Breiling are with the Fraunhofer Institute for Integrated Circuits IIS, Am Wolfsmantel 3, 91058 Erlangen, Germany. Ferdinand Pscheidl is with the Fraunhofer Research Institution for Microsystems and Solid State Technologies EMFT, Hansastraße 27d, 80686 Munich, Germany.

I. INTRODUCTION

THE rapid adoption of deep learning in recent years has led to a growing demand for efficient AI hardware accelerators. In particular, edge applications with very specific and stringent KPI requirements, i.e. in terms of power consumption or latency, stand to benefit from highly customized solutions. However, designing a custom inference accelerator for edge AI applications requires a multi-disciplinary hardware/software co-design approach that combines neural network algorithm design and software development with integrated circuit design. To reach the targeted KPIs within a short time-to-market despite the complexity of this task, dedicated tools and workflows for optimizing and verifying the hardware design are mandatory. Taken together, such tools form a hardware/software co-design tool chain that, we argue, is crucial to make the development of custom neuromorphic hardware accelerators viable.

In this paper, we present an example of such a tool chain, developed in the AN-DANTE project, for a mixed-signal inference accelerator with analog In-Memory Computing (IMC) and explain the rationale behind it.

The paper is organized as follows: after providing a brief overview of the entire tool chain in section II, we explain each of its components in section III, followed by concluding remarks in section IV.

II. TOOLCHAIN OVERVIEW FOR NEUROMORPHIC COMPUTING

The design of an inference accelerator with high performance, e.g. in terms of its energy efficiency, latency and throughput, requires more than good circuit design. It also requires tools that minimize the memory footprint, data movement and access, and that can provide (optimized) sets of instructions for the specific hardware. To form a coherent tool chain, these tools need to be mutually compatible, and they need to be designed with the target hardware in mind.

Fig. 1 shows the different components of one of the toolchains used in the ANDANTE project for the design of optimized edge-AI inference accelerators. It comprises five specific tools, each ex-

plained in the following section, that take as input the hardware building blocks from which the system is constructed, neural network architectures and the labeled data to train them, as well as a description of the hardware system architecture and its limitations. As output, these tools provide (part of) the hardware design, the neural networks and instructions to deploy on it, as well as estimations of its KPIs.

The Hardware-Aware Training tool takes the labeled data set and the NN model as input and generates a quantized model. In this way, the quantized model takes into consideration the quantization of the NN parameters and some errors induced by the hardware components. Afterwards, the quantized NN model is provided to the Mapper & Compiler tool to obtain the compiled program that will be used by both the Architecture Exploration and Simulation tool and the Deployment & Runtime API. The NN Hardware Generator Tool automates partially the generation of the custom mixed-signal inference accelerator.

State-of-the-art design and simulation environments for neural networks like N2D2 [1] allow quantization-aware training but do not provide the possibility to train the network with the weight variations introduced by the IMC analog circuits. Already existing state-of-the-art mapping tools for digital accelerators like Timeloop can also be used to represent mixed signal inference accelerators containing simple analog crossbar arrays [2]. However, complex constraints related to dataflow, computational resources and scheduling of standalone accelerators cannot be integrated in Timeloop. As the existing tools are not able to provide enough functionality and flexibility to incorporate the restrictions and artifacts produced by an analog neural network accelerator, a new framework of tools

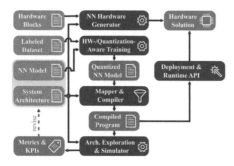

Fig. 1. Toolchain for the design of edge-AI mixed-signal inference accelerators.

dedicated to such circuits has to be developed.

III. DESCRIPTIONS OF THE INDIVIDUAL TOOLS

A. Hardware –Aware Training

Hardware-Aware Training (HAT) consist of two main components: Quantization-Aware Training (QAT) and Fault-Aware Training (FAT). In this section, we explain why both are required and present our approach of constructing the NNs and training them.

Low-power consumption NNs heavily rely on reducing the processing power of the network as they are deployed at the edge with limited power budget. It has been shown that limiting the NN parameters and arithmetic to fixed-point has a big impact on reducing the energy consumption, as it allows a smaller memory footprint of the neural network with less memory access and smaller multipliers [3] [4] [5]. This approach requires quantizing the weights, activation and bias parameters with different bit width and fixed-point format. Generally, there are two main approaches for NN quantization: I) quantizing a pre-trained NN (Post-Training Quantization) and II) including the quantization in training (QAT). The first works well when the NN is quantized to 16 or (in some cases) 8 bits while the latter is proven to achieve better performance (accuracy) when severe quantization is required (e.g. 1 up to 4 bits) [5] [6] [7].

Additionally, non-ideal circuit behaviors (errors) can degrade the performance of the NN. The HW errors can be mitigated by modelling and injecting them in SW during training. This has been shown to produce more reliable NNs, since they are robust against the errors, they were trained with [9]. This is referred to as FAT. A HAT tool was implemented including both QAT and FAT.

The HAT tool uses the QAT built in Xilinx Brevitas [8], and extends it to include FAT. A *Fault-Aware Quantizer (FAQ)* is implemented including both Quantization (Q) and Fault injection before (*PreQ Op*) and after (*PostQ Op*) quantization operation, see Fig. 2. Any quantizer (FAQ or the standard Brevitas Q) or no quantizer (NoQ) can be used with weights, inputs, bias and/or outputs of any layer (convolutional, fully connected, activation, pooling, etc.), see Fig. 3. For example, we could quantize some inputs with the FAQ and others with the standard Brevitas Q.

Moreover, the faults/errors that occur in hardware can be injected in two places: before (*PreQ* Op) or after (*PostQ* Op) the quantization (Q) takes places. PreQ Op and *PostQ Op* are modular and can model any error associated with a hardware implementation. Currently, the HAT implements relative noise (1), absolute noise (2), scaling (3) and bit flips.

$$x = x + x \cdot n. \quad (1)$$
$$x = x + n. \quad (2)$$
$$x = x \cdot s. \quad (3)$$

where x is the input value, n is the noise and s is the scaling factor.

However, any other error source could be modeled and added easily. In this way, the HAT tool can be used for different hardware implementations that may require different errors/variation implementations.

Here, we list few examples where errors can be injected/modeled with *PreQ Op* or *PostQ Op* of the *FAQ*.

Fig. 2. Fault aware quantizer (FAQ) implementation.

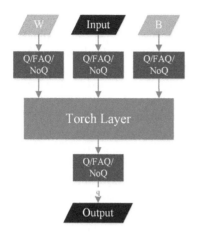

Fig. 3. Block diagram of a Hardware-Aware Training layer.

1) In analog crossbars, where the synaptic weights are implemented with memristors, the nominal value of these memristors may vary in the manufacturing process. This error can be modeled as noise in the *PostQ Op* of *FAQ* at the weight.
2) The electrical noise affects the input signal (partial sum) before applying the ADC (the quantizer in this analogy). This error can be modeled as noise in the *PreQ Op* of the FAQ at the layer output.
3) Stuck at 0 and stuck at 1 errors [9] result in a processing element being stuck at one value. This can be modeled by fixing the values in the PostQ Op of the FAQ at the layer output..
4) Bit flips can be modeled in the PostQ Op of the FAQ at the layer output or input .

B. Mapper & Compiler

Convolutional Neural Networks (CNNs) are widely used in computer vision and audio applications. CNNs consist of different layers: Convolutional, Fully Connected, Pooling, Padding, and Normalization layers. Padding and Pooling layers add pre and post-processing of the data on the chip. The main operation to be executed by an accelerator are multiply-accumulate (MAC) operations. With increasing size of neural networks and the requirement to reduce the energy consumption and increase the throughput, the execution of the MAC operations on the targeted hardware accelerator has to be optimized which can be achieved by optimizing the data movement. The use of analog crossbar arrays has surpassed the energy efficiency achieved by the digital processing accelerators for computing the MAC operations [10] [11]. To make an efficient inference on the accelerator, the compiler attempts to allocate adequate on-chip resources for the processing of NN layers and schedules the instructions accordingly.

In mixed-signal inference accelerators, a mapper and compiler process quantized neural network models, stored in an Intermediate Representation (IR) format like ONNX format, and the architecture specification. The Mapper explores the mapping design space and sends the valid mappings to the Analyzer, see Fig. 4, following a pre-defined set of rules:

1) Realize the critical constraints (like computational elements, ADC bitwidth and its availability to the processing core, stationarity of dimension) and exploit the hardware limitations

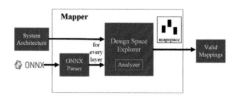

Fig. 4. Overview of the Mapper Tool.

2) For mixed-signal inference accelerators based on a crossbar architecture, maintain stationarity of weights. If required, include reconfiguration of weights and try to minimize the reloading of weights from the global buffer.
3) Explore parallel computation of outputs on the hardware to reduce latency.
4) Data fetch/write and processing in CWHN (first #channels, #width, #height, #batch of filters) format.

The mapper takes into account constraints like neural network layer dimensions, limited on-chip resources like memory sizes, computational block dimensions, and Network-on-Chip (NoC) Control that induce numerous ways for mapping the same network layer on the same hardware. The Analyzer within the Design Space Explorer discards such invalid mappings that cannot satisfy physical limitations of the hardware while valid mappings are translated to an intermediate mapping output format. The outcome of these different mapping possibilities will vary in terms of energy consumption and latency.

Fig. 5 shows an overview of a compiler tool and its sub-modules for a mixed-signal inference accelerator with analog in-memory computing. The Evaluator evaluates the valid mappings based on the performance heuristic expected (currently, max usage of on-chip computational resources). Following this, the Scheduler converts the mapping into appropriate instructions. For standalone accelerators with custom instruction sets, a dedicated compiler needs to be developed. The intermediate mapping output generated by the tool can be used to reduce the exploration time by any compiler that could translate the mapping information. Further, the compiler can pipeline the layers to process the compu-

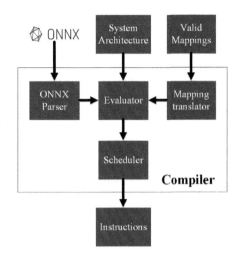

Fig. 5. Overview of the Compiler Tool.

tations in parallel. For example: as soon as layer *n-1* generates sufficient outputs to start processing layer *n*, the output buffer of layer *n-1* sends the inputs to the on-chip locations where layer *n* will be processed.

In summary, the goal of such a Mapper & Compiler tool is to efficiently explore the design space and ensure the best possible mapping for the accelerator.

C. Architecture Exploration & Simulation

In order to provide scalability with respect to smaller or larger NNs for various use cases, mixed-signal accelerators combine analog in-memory computing with a multi-core system architecture and distributed memory. The design complexity of such SoCs, e.g. in terms of the number of transistors and other digital or analog components, rules out simulations at transistor level or RTL simulations of the entire system (i.e. at top-level) with realistic use-cases due to the simulation times needed on state-of-the-art simulation clusters. Therefore, a functional model of the SoC with decreased simulation time will make automated architectural design space exploration feasible. Also, it will enable the optimization of the aforementioned hardware KPIs, system level simulations and the validation of the Mapper & Compiler output at an early stage.

An architecture exploration tool allows to optimize the system architecture for a given use case or, conversely, optimize a NN under the constraints imposed by the system architecture, while estimating the relevant KPIs for a given configuration of the hardware architecture and NN model. The performance of such complex systems-on-chips (SoCs) can be measured in terms of use-case specific key performance indicators (KPIs) like energy consumption, latency per inference and inference accuracy [12]. The KPIs depend on many aspects of the architecture, e.g. memory type and sizes, number of processing units [13] [14], bit accuracy, communication bandwidths, technology node, as well as hardware induced inaccuracies like noise or quantization errors [15]. Even though the hardware-aware training (HAT) tool already estimates some of these KPIs, multiple hardware effects – like splitting up the calculation of a layer into multiple processing cores – are not taken into consideration.

Therefore, a functional model of the SoC sacrifices accuracy in favor of speed by transitioning from transistor or gate level models to more abstract models; e.g. pin and cycle accurate at the module interfaces or even more abstract transaction level models (TLM).

Besides simulation time, having a functional model makes possible to carry out an architecture exploration, design and verification workflow already at an early stage of development. This greatly simplifies hardware and software stack verification (e.g. verifying the compiler output) and reduces the time to market for the developed complex heterogeneous SoCs that require close HW/SW co-design.

To ensure that the system level simulation and verification of such an abstract model is accurate, it needs to be complemented by and compared against more detailed models of the individual modules at various lower levels of abstraction. The right level of detail can also vary from module to module; for example, while the internal timing of some processor modules may be safely abstracted, the timing of other components such as network-on-chip (NoC) and bus arbitration circuits may be critical for the operation of the system (e.g. contention), and, therefore, have to be modeled cycle-accurately. This can be especially challenging during early development, where hardware specifications are not yet stable, and any changes need to be continuously integrated throughout these multiple levels.

Several tools exist to address this problem of modelling complex systems across multiple levels of abstraction; similar to prior work [16], the tool chain described here employs SystemC [17], [18], which can be used with commercial simulation and verification tools like Incisive (from Cadence Design Systems) [19] and ModelSim (from Siemens) [20].

In the ANDANTE project, a functional System C model of one of the mixed-signal inference accelerators has been developed and can efficiently simulate the system's behavior at a higher level of abstraction, while its correctness can be verified by lower level hardware simulations of its individual parts. This abstract model is used to test and verify the designed hardware, to estimate the KPIs (like accuracy, latency, throughput and energy consumption), and to make informed architecture and design choices in order to optimize these KPIs (e.g. selecting the ADC bit resolution).

D. Deployment & Runtime API

Before the hardware can be deployed in production, various tests need to be performed to verify the correct operation of the hardware and to refine the estimated KPIs with real-world measurements. Therefore, we designed a broad suite of automated measurements and tests that compare the ground-truth data produced by hardware measurements against the expected outputs provided by the simulation tools. To support such automated testing, the hardware must be designed accordingly from early on ("design-for-testing").

To finally deploy the developed hardware in production, simplified, user-friendly APIs are needed to communicate parameters, data and results between the chip and the host hardware. Depending on the host systems for a given application, such an API may need to support microprocessors, e.g. via a C library, and/or conventional computers, e.g. via a python library. In the ANDANTE project, our work has so far focused on APIs for testing, but an API for deployment outside of the lab is planned for future work.

E. Neural Network Hardware Generator

To execute neural networks efficiently in terms of energy consumption, latency and accuracy, dedicated hardware is needed especially in edge applications. Further, the configuration of the circuits must be set according to the requirements of the neural network to execute, which includes for example the maximum kernel size of a convolutional layer that the hardware can execute, how many neurons can be computed in parallel or the number of quantization levels of the weights. Analog circuits, dedicated to be used in neuromorphic hardware, consist of the assembly of thousands of the same unit circuits, like synapses and neurons, which have to be placed and routed. With each change of one or multiple of the available hardware parameters, the circuits must be redone which leads to a long time-to-market, especially if an analog-mixed signal accelerator should be implemented, since only minimal automation is available for analog circuit design. Therefore, a hardware generator tool for the automatic generation is required for the circuit design of neural network accelerators.

The Neural Network Hardware Generator tools implement the automation of the circuit design for dedicated Neural Network accelerators. The tool takes the architecture specifications like the number of weights and neurons as well as block level implementations as input and assembles them into the final accelerator circuit. The block level circuits are synapse and neuron circuits together with their required framework circuits. According to the architecture specifications, the tool then instantiates and connects these block level modules to finalize the accelerators circuits.

Within the ANDANTE project, the functionality of our Neural Network Hardware Generator Tool for a multi-purpose accelerator architecture with analog-mixed signal data processing based on the Fraunhofer IIS internal automation framework called UnilibPlus [21] was extended. Fig. 6 shows the hierarchy of the designed circuit, where the green blocks are manually designed block level circuits, whilst the blue blocks are automatically created by the tool. Thus, the analog core is automatically generated, at schematic and layout levels, from the *Crossbar Array* and *ADC Array* cells. The Neural Network

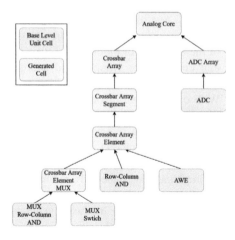

Fig. 6. Circuit Hierarchy created by Neural Network Hardware Generator.

Hardware Generator Tool takes the user created *ADC* unit cell, places them together, routes the inputs to the correct place where they should be connected to the *Crossbar Array* and executes the power routing and finally interconnects the ADCs generating then the *ADC Array* cell. The *Crossbar Array* cell is generated from *Crossbar Array Segment* cells. These *Crossbar Array Segments* contain the *Crossbar Array Element* which in turn are generated from base level unit cells, with in this case are the synapse circuit (*AWE*), the *Row* and *Column* selection logic (*Row-Column AND*) and the generated multiplexer cells, which are used to interconnect the *Crossbar Array Elements* to each other and to the *ADC Array*. Moreover, the hierarchy shown in Fig. 6 can be changed in order to explore different architectures of the analog core, while using the same unit cells, to examine possible reductions in area and to provide options for tradeoffs between performance, leakage and energy consumption.

This tool does not only reduce the design time and therefore the time-to-market but also enhances the stability of the design process. Neural network accelerators are immensely complex and large circuits, which leads to a high possibility of errors if the design is created manually. By automating the design process, examples can be created, simulated and verified to ensure a correct implementation. Any up or downscaled version of these exemplary circuit is then considered to be correct since the process for implementing them is executed by the same algorithm. Therefore, the possibility of errors in the design is reduced. Furthermore, it enables the parallel development of neural network algorithm and the neural network circuits as many different versions of the circuits and algorithm can be created and tested within a reasonable time frame.

IV. CONCLUSION AND FUTURE WORK

The development and deployment of custom AI accelerators is a multidisciplinary effort that should be addressed from a holistic system's perspective. In particular, careful co-design of the neural networks, soft- and hardware is required to achieve a good utilization of the hardware in practice. In the ANDANTE project, we developed a stack of tools to support this designflow, from the assessment of relevant KPIs, to hardware-aware training of neural networks, to simulators, compilers and drivers for the hardware, all the way down to (partially) automated generation of the hardware itself. We argue that this approach, laborious as it may be, is highly beneficial, if not necessary, for the design of complex AI accelerators.

To facilitate similar AI accelerator design efforts in the future, further work should be invested to integrate these and/or similar tools into a cohesive and general framework. Moreover, such framework is necessary for the benchmarking of AI accelerators based on use case requirements.

REFERENCES

[1] "CEA-LIST / N2D2". [Online]. Available: https://github.com/CEA-LIST/N2D2

[2] A. Parashar et. al., "Timeloop: A systematic approach to dnn accelerator evaluation," in *2019 IEEE international symposium on performance analysis of systems and software (ISPASS)*, IEEE, 2019, pp. 304-315

[3] B. Moons, K. Goetschalckx, N. Van Berckelaer and M. Verhelst, "Minimum energy quantized neural networks," in *Proc. 2017 51st Asilomar Conference on Signals, Systems, and Computers*, 2017, pp. 1921-1925, doi: 10.1109/ACSSC.2017.8335699.

[4] J. Johnson, "Rethinking floating point for deep learning,". *arXiv preprint arXiv:1811.01721,* 2018.

[5] Q. Ducasse, P. Cotret, L. Lagadec and R. Stewart,. "Benchmarking Quantized Neural Networks on FPGAs with FINN," *arXiv preprint arXiv:2102.01341*, 2021.

[6] M. Courbariaux, Y. Bengio and J.P. David, "Binaryconnect: Training deep neural networks with binary weights during propagations," *Advances in neural information processing systems*, 28, pp. 3123-3131, Nov. 2015.

[7] M. Rastegari, V. Ordonez, J. Redmon and A. Farhadi, "Xnor-net: Imagenet classification using binary convolutional neural networks,"in *European conference on computer vision,* Springer, Cham., 2016, pp. 525-542.

[8] A. Pappalardo, G. Franco, and nickfraser, "Xilinx/brevitas: Cnv test reference vectors r0," May 2020. [Online]. Available: https://doi.org/10.5281/zenodo.3824904.

[9] U. Zahid, G. Gambardella, N. J. Fraser, M. Blott, and K. Vissers, "FAT: Training Neural Networks for Reliable Inference Under Hardware Faults," in *2020 IEEE International Test Conference (ITC)*, Nov. 2020, pp. 1–10, doi: 10.1109/ITC44778.2020.9325249.

[10] T. Luo, S. Liu, L. Li, Y. Wang, S. Zhang, T. Chen, Z. Xu, O. Temam and Y. Chen, *"DaDianNao: A Neural Network Supercomputer,"* *IEEE Transactions on Computers,* 66, pp.73-88, 2017.

[11] A. Shafiee, A. Nag, N. Muralimanohar, R. Balasubramonian, J. P. Strachan, M. Hu, R. S. Williams, and V. Srikumar, "ISAAC: a convolutional neural network accelerator with in-situ analog arithmetic in crossbars," in *Proceedings of the 43rd International Symposium on Computer Architecture (ISCA '16,* IEEE Press, pp. 14–26, 2016, https://doi.org/10.1109/ISCA.2016.12.

[12] S. Narduzzi, L. Mateu, P. Jokic, E. Azarkhish and A. Dunbar, "Benchmarking Neuromorphic Computing for Inference," in *Industrial Artificial Intelligence Technologies and Applications*, River Publishers, 2022, pp. 1-16.

[13] X. Peng, S. Huang, H. Jiang, A. Lu, and S. Yu, "DNN+NeuroSim V2.0: An End-to-End Benchmarking Framework for Compute-in-Memory Accelerators for On-Chip Training," *IEEE Transactions on Computer-Aided Design of Integrated Circuits and Systems*, vol. 40, no. 11, pp. 2306–2319, Nov. 2021, doi: 10.1109/TCAD.2020.3043731.

[14] Y. N. Wu, J. S. Emer, and V. Sze, "Accelergy: An Architecture-Level Energy Estimation Methodology for Accelerator Designs," in

2019 IEEE/ACM International Conference on Computer-Aided Design (ICCAD), Nov. 2019, pp. 1–8, doi: 10.1109/ICCAD45719.2019.8942149.

[15] L. Mei, P. Houshmand, V. Jain, S. Giraldo, and M. Verhelst, "ZigZag: Enlarging Joint Architecture-Mapping Design Space Exploration for DNN Accelerators," *IEEE Transactions on Computers*, vol. 70, no. 8, pp. 1160–1174, Aug. 2021, doi: 10.1109/TC.2021.3059962.

[16] D. Bortolotti, C. Pinto, A. Marongiu, M. Ruggiero, and L. Benini, "VirtualSoC: A Full-System Simulation Environment for Massively Parallel Heterogeneous System-on-Chip," in *2013 IEEE International Symposium on Parallel & Distributed Processing, Workshops and Phd Forum*, Mai 2013, pp.. 2182–2187, doi: 10.1109/IPDPSW.2013.177.

[17] P. R. Panda, "SystemC - a modeling platform supporting multiple design abstractions," in *International Symposium on System Synthesis (IEEE Cat. No.01EX526)*, Sep. 2001, pp. 75–80, doi: 10.1145/500001.500018.

[18] G. Arnout, "SystemC standard," in *Proceedings 2000. Design Automation Conference (IEEE Cat. No.00CH37106)*, Jan. 2000, pp. 573–577. doi: 10.1109/ASPDAC.2000.835166.

[19] "Computational Software for Intelligent System DesignTM". Cadence. [Online]. Available: https://www.cadence.com/en_US/home.html

[20] "ModelSim HDL simulator". Siemens. [Online]. Available: https://eda.sw.siemens.com/en-US/ic/modelsim/

[21] R. Müller, et al., "Hardware/Software Co-Design of an Automatically Generated Analog NN," in *International Conference on Embedded Computer Systems*, 2022, pp. 385-400, Springer, Cham.

Loreto Mateu obtained her B.S. in Industrial Engineering in 1999, her M.S. in Electronic Engineering in 2002 and her PhD degree (with highest honors) in June 2009 with a thesis titled Energy Harvesting from Human Passive Power at the Universitat Politècnica de Catalunya.

In June 2007, she joined Fraunhofer IIS as research engineer and became chief scientist in 2012. Since 2018 she is group manager of the Advanced Analog Circuits group. Her research interests include ultra-low power design, AC-DC and DC-DC converters, neuromorphic hardware and energy harvesting.

Johannes Leugering is a researcher in the area of neuromorphic computing at the Fraunhofer Institute for Integrated Circuits (IIS). He received a PhD in computational neuroscience (Dr.rer.nat, with highest honors) from Osnabrück University on the topic of *"Neural mechanisms of information processing and transmission"* in 2021.

Since 2019, he has been working at Fraunhofer IIS as an expert for neuromorphic computing concepts and architectures, particularly in the context of spiking neural networks. Since 2022, he is chief scientist in the Broadband and Broadcast department.

Roland Müller born on the 19. January 1994 obtained his bis B.Eng. at the OTH Regensburg in 2017 and his M.Sc. in 2019 at the FAU Erlangen, both in electrical engineering.

In May 2019, he joined the department of Integrated Circuits and Systems at Fraunhofer IIS, Erlangen (Germany), where he is working in the field of analog-mixed signal design of neural network accelerators and design automation for such circuits. Currently, he is pursuing his PhD.

Yogesh Patil completed his M.E. in Information Technology in 2021 with thesis titled, "Design Space Exploration of Neural Network Hardware Architectures", from SRH University Heidelberg/Germany and B.E. in Electronics and Telecommunications from Savitribai Phule Pune University (Formerly University of Pune), India in 2018.

Since January 2021, he has been contributing to the research in neural network accelerators with In-Memory Computing architectures at Fraunhofer Institute for Integrated Circuits (Fraunhofer IIS, Germany).

Maen Mallah is a researcher in the area of embedded AI at the Fraunhofer Institute for Integrated Circuits (IIS). He obtained his B.Sc in Telecommunication Engineering in 2014 from An-Najah National University, Palestine and his M.Sc in Electrical Engineering in 2018 from Bilkent University, Turkey with a thesis titled "Multiplication Free Neural Networks".

In March 2018, he joined Fraunhofer IIS. His main work and interest focuses on implementing and optimizing NNs for Edge applications and designing the special SW tools required for such a task with a special focus on Quantization- and Fault-aware training.

Marco Breiling conducted studies at the Universität Karlsruhe/Germany (now Karlsruhe Institute of Technology – KIT), the Norges Tekniske Høgskole (NTH) (now Norges Teknisk-Naturvitenskapelige Universitet – NTNU), the Ecole Supérieure d'Ingénieurs en Electronique et Electrotechnique (ESIEE) and the University of Southampton, and graduated with a Dipl.-Ing. (equivalent to master's) degree from KIT in 1997. He earned his PhD degree (with highest honors) in digital communications from Universität Erlangen/Germany in 2002.

Since 2001, he has been working at Fraunhofer IIS in the field of signal processing, digital communications and digital design. He held the chief scientist position of the Broadband & Broadcast department from 2013 until 2021. Moreover, he is a Distinguished Lecturer of the IEEE Broadcast Technology Society.

Ferdinand Pscheidl obtained his B.S. in Bachelor of Science in electrical engineering at Technische Universität München in 2018. In 2021 he received his Master of Science in electrical engineering at Technische Universität München with focus on circuit design and machine learning.

In 2021 he joined Fraunhofer EMFT as research engineer. His research interests include ultra-low power design, neuromorphic hardware, software and development tools.

Low-Power Vertically Stacked One Time Programmable Multi-bit IGZO-Based BEOL Compatible Ferroelectric TFT Memory Devices with Lifelong Retention for Monolithic 3D-Inference Engine Applications

Sourav De, Sunanda Thunder, David Lehninger, Michael P.M. Jank,
Maximilian Lederer, Yannick Raffel, Konrad Seidel, and Thomas Kämpfe

Abstract—This article demonstrates indium gallium zinc oxide-based onetime programmable ferroelectric memory devices with multilevel coding and lifelong retention capability. The entire integration process was conducted in the back-end-of-line with a maximum process temperature of 350°C. The fabricated devices demonstrate data retention up to 10^4 seconds, which was used to estimate the retention property up to 10^8 seconds. We observed a marginal drop in channel current after 10^8 seconds, which makes them suitable for inference engine application. The compatibility with the back-end-of-line process enables monolithic 3D integration of the devices with standard technology. We have evaluated the performance of this indium gallium zinc oxide-based onetime programmable ferroelectric thin film transistor for inference engine applications. The system-level simulation was performed to gauge the performance of the devices as synapses in multilevel perceptron-based neural networks. The synaptic devices could achieve 97% for inference-only applications with MNIST data. The accuracy degradation was also limited to 1.5% over 10 years without retraining. The proposed inference engine also showed superior energy efficiency and cell area of 95.33 TOPS/W (binary) and $16F^2$, respectively.

Index Terms—Hafnium Oxide, IGZO, FeTFT, non-volatile memory, variation, neural networks.

Sourav De, Sunanda Thunder, David Lehninger, Maximilian Lederer, Yannick Raffel, Konrad Seidel, and Thomas Kämpfe are associated with Fraunhofer-Institut für Photonische Mikrosysteme IPMS, Center Nanoelectronic Technologies, Dresden, Germany. Michael P.M. Jank is associated with Fraunhofer-Institut für Integrierte Systeme und Bauelementetechnologie. Erlangen, Germany.

This research was partly funded by the ECSEL Joint Undertaking project ANDANTE in collaboration with the European Union's Horizon 2020 Framework Program for Research and Innovation (H2020/2014-2020) and partly by the European Union's ECSEL Joint Undertaking under grant agreement n° 826655 project TEMPO (Corresponding author:Sourav De). Email:Sourav.de@ipms.fraunhofer.de

I. INTRODUCTION

RECENT developments in the research of hafnium oxide(HfO_2) based ferroelectric memories in the form of ferroelectric (Fe) field effect transistors (Fe-finFETs) [1–7], Fe-finFETs [8–12] and ferroelectric thin film transistors (FeTFT) [13–17] have paved the way for further scaling, 3D-integration, and system level application of Fe memory devices. So far, most of the applications of FeTFT have focused on improving endurance characteristics, which is crucial for online training. However, the online training of the neural network requires endurance above 10^8 cycles [8, 9], which limits the retraining of the neural network if necessary. Therefore, in this work, we primarily focus on the offline training of the neural network with the inference-only operation for hardware. While the endurance characteristics are essential for online training of the neural network, retention is vital for conducting the inference-only operation. The depolarization field across the ferroelectric layer plays a vital role in retention characteristics.

The degradation in data retention over time necessitates retraining the neural network after a certain amount of time and renders multi-bit per cell operation futile for long-term applications.

In this work, we demonstrate indium gallium zinc oxide-based (IGZO) one-time programmable (OTP) memory devices with lifelong retention for inference engine applications. The IGZO-based FeTFT device was fabricated using a gate-first process with different ratios of the ferroelectric layer and channel layer thickness. The data retention capability is related to the depolarization field across the ferroelectric layer. The primary motive of this work was to reduce the depolarization field without affecting the memory window. It is pretty well known that the carrier's mobility and the relative thickness of the dielectric and semiconductor layer regulates the voltage drop across the ferroelectric layer

and channel. It is worth noting that the mobility of electrons in IGZO is much higher than the mobility of the holes. ?We observed that while $t_{IGZO} > \frac{1}{2} t_{HZO}$, most of the electric field (ξ) is dropped across the IGZO film during erase operation. Therefore, the impact of the depolarization field is also lowered across the HZO film, simultaneously facilitating the fabricated devices' onetime programming (OTP) and lifelong retention capability. The IGZO-based OTP synaptic devices occupy a cell area of $16F^2$, where F is the lithography feature size. The devices also demonstrate lifelong retention, which is the most critical characteristic for inference application. IGZO-based TFTs demonstrate 2bits/cell operation, with extrapolated retention for each state above ten years. We have further evaluated the IGZO-TFT's performance as a synaptic inference engine device. The performance of these synaptic devices in terms of area and energy efficiency demonstrates their suitability of this device for in-memory-computing (IMC) applications. The synaptic devices maintained an inference accuracy above 95% for ten years with a multi-layer perceptron (MLP) based neural network (NN) for MNIST dataset recognition without retraining.

II. EXPERIMENTS

We began our experiment by fabricating metal-ferroelectric metal (MFM) and metal-semiconductor-ferroelectric-metal (MSFM) capacitors on highly doped (boron) 300mm silicon wafers using industry-standard production tools. The titanium nitride (TiN) bottom electrode was deposited via atomic layer deposition (ALD). Titanium tetrachloride (TiCl$_4$) and ammonia (NH$_3$) were precursors for the ALD process. The ferroelectric layer (Hf$_x$Zr$_{1-x}$O$_2$) was deposited at

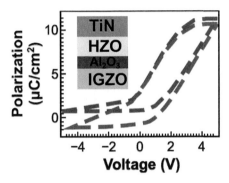

Fig. 1. The P-V response of HZO-based ferro-electric capacitors as IGZO as the bottom electrode shows asymmetric swing with negligible negative switching. The absence of negative polarizations plays an essential role in facilitating OTP features in Fe-TFTs.

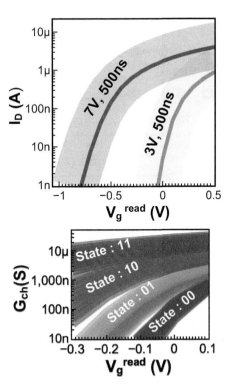

Fig. 2. (a). WRITE operation in IGZO based Fe-TFTs with 500ns wide pulses of amplitude 3V and 7V. (b). 2bits/cell operation for IGZO-based Fe-TFT OTP devices.

300°C by ALD. With $HfCl_4$ and $ZrCl_4$ precursors, H_2O as an oxidizing agent, and Ar as a purging gas. For MSFM capacitors, the IGZO was deposited by RF magnetron sputtering. The thickness of the IGZO varied between 5nm and 30nm. A 2nm thick layer of Al_2O_3, deposited by ALD, was used as an interfacial layer between $Hf_xZr_{1-x}O_2$ and IGZO. The TiN top electrode was deposited by magnetron sputtering, where the deposition temperature is below 100°C. The annealing for crystallization was carried out at 350°C. The FE-TFTs were fabricated on standard silicon wafers. 100nm SiO_2 was used to insulate the devices from the substrate. 50nm of TiN was deposited by ALD and patterned via e-beam lithography and reactive ion etching to form bottom gate electrodes. ALD deposited HZO of 10nm thickness, followed by 2nm ALD Al_2O_3. Finally, the devices were annealed in air at a temperature of 350°C for 1h.

The polarization versus field (P–E) measurements were performed with a triangular waveform at a frequency of 1kHz (Fig. 1). The formation of both accumulation and inversion layers are the basic requirements for conducting the program and erasing the FE-TFTs. The mobility of the electrons ($\mu_n \approx 10$ cm^2 V^{-1}s^{-1}) in IGZO-based semiconductors is high. Therefore, the accumulation layer in n-type IGZO devices is formed within a short time, 100ns. Contradictorily, the hole mobility (μ_p) is deficient, and the inversion layer formation is complex when most of the electric field is dropped across IGZO. Proper tuning of the thickness of IGZO and HZO resulted in the omission of erasing capability in the fabricated TFT devices. We have observed that an inversion channel is not formed even after applying long pulses (2s) of amplitude up to -6V. The dielectric breakdown happens before the formation of

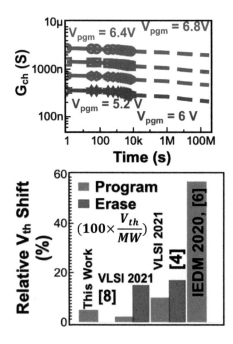

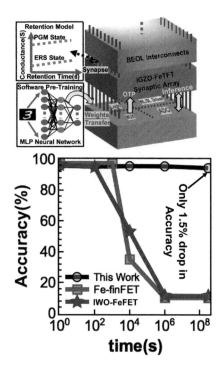

Fig. 3. (a). The measured retention characteristics show stable retention of 4-states for ten years without any loss. (b). Benchmarking the retention performance relative V_{th} shift w.r.t MW proves that IGZO-based OTP devices have maximum long-term data retention capability.

Fig. 4. (a). The modus operandi of MLP NN. (b). The reported inference engine shows life-long lossless inference operation.

the inversion layer. This invokes the OTP scheme in IGZO-TFTs, which is also responsible for lifelong data retention capability. The binary and 2bits/cell READ-WRITE operations are demonstrated in Fig. 2 (a,b).

The retention characteristic (Fig.3 (a)) shows only slightly conductance degradation after 10 years, which is superior to other state-of-the-art Fe-FETs. In Fig. 3(b), we showed the lowest V_{th} shift w.r.t memory window (MW) compared to other state-of-art Fe-FET devices.

Finally, the system-level validation for the inference-only application was performed using the CIMulator platform [7]. The modus-operandi for the IGZO-FeTFT inference engine is described in Fig. 4(a). The multi-layer perceptron-based neural network (MLP-NN) has

three layers, including 400 input-layer nodes, 100 hidden-layer nodes, and 10 nodes in the output layer. After completing the training, the synaptic weights were written to the FeTFT-based synaptic core. The inference task is performed once the synaptic weights are updated in the FeFET devices. The measured 2bits/cell operation with experimentally calibrated variations was used while conducting the neuromorphic simulations. Due to retention degradation, the accuracy drop was limited to only 1.5% over 10 years, while other state-of-the-art ferroelectric memory devices could only retain 11% inference accuracy after 10 years. We have further simulated the impact of ADC precision in the inference engine, which shows that using area efficient 1bit ADCs can boost the system performance in terms of area and energy efficiency with ?only accuracy drops of 1.87% and maintains the inference accuracy above 94% for 10 years. Table I. summarizes the performance of this device w.r.t other state-of-the-art devices.

TABLE I
BENCHMARKING

Device Type	Fe-FinFET [9]	IWO-FeFET [14]	This Work
M3D Integrator	No	Yes	Yes
Cell Area (F^2)	15F^2	15F^2	16F^2
R$_{on}$ (Ω)	100K	4M	100M
MW @10 years	1	0.2	1
Inference Accuracy Drop @10 years	~85%	85%	1.5%
Energy Efficiency (TOPS/W)	N/A	71.04	95.33 (Binary)

III. IV. CONCLUSIONS

Ultra-low power multi-bit IGZO-based OTP Fe-TFTs with lifelong retention have been fabricated with a maximum process temperature of 350°C. The devices demonstrate 2bits/cell operation. Long-term non-disturbing retention property makes these devices.

ACKNOWLEDGEMENT

This research was funded by the ECSEL Joint Undertaking project ANDANTE in collaboration with the European Union's Horizon 2020 Framework Program for Research and Innovation (H2020/2014-2020) and National Authorities under Grant No. 876925. We thank Dr. Hoang-Hiep Le and Prof. Darsen Lu from National Cheng Kung University for helping us with the neuromorphic simulations.

REFERENCES

[1] T. Ali *et al.*,"A Multilevel FeFET Memory Device based on Laminated HSO and HZO Ferroelectric Layers for High-Density Storage," *2019 IEEE International Electron Devices Meeting (IEDM)*, 2019, pp. 28.7.1-28.7.4, doi: https://doi.org/10.1109/IEDM19573.2019.8993642

[2] Kim, M.-K., & Lee, J.-S. Ferroelectric Analog Synaptic Transistors. *Nano Letters*, 19(3), 2044–2050.2019 https://doi.org/10.1021/acs.nanolett.9b00180.

[3] Kim, D., Jeon, Y.-R., Ku, B., Chung, C., Kim, T. H., Yang, S., Won, U., Jeong, T., & Choi, C. (2021). Analog Synaptic Transistor with Al-Doped HfO 2 Ferroelectric Thin Film. *ACS Applied Materials & Interfaces*, 13(44), 52743–52753. https://doi.org/10.1021/acsami.1c12735

[4] Soliman, T., Gmbh, R. B., Laleni, N., & Ipms, F. (2022). FELIX: A Ferroelectric FET Based Low Power Mixed-Signal In-Memory Architecture for DNN Acceleration. *ACM Transactions on Embedded Computing Systems.* https://doi.org/https://doi.org/10.1145/3529760

[5] Trentzsch, M., Flachowsky, S., Richter, R., Paul, J., Reimer, B., Utess, D., Jansen, S., Mulaosmanovic, H., Muller, S., Slesazeck, S., Ocker, J., Noack, M., Muller, J., Polakowski, P., Schreiter, J., Beyer, S., Mikolajick, T., & Rice, B. (2017). A 28nm HKMG super low power embedded NVM technology based on ferroelectric FETs. *Technical Digest - International Electron Devices Meeting, IEDM.* https://doi.org/10.1109/IEDM.2016.7838397

[6] Wang, P., & Yu, S. (2020). Ferroelectric devices and circuits for neuro- inspired computing. *MRS Communications*, 10(4). https://doi.org/10.1557/mrc.2020.71

[7] S. De et al., "READ-Optimized 28nm HKMG Multibit FeFET Synapses for Inference-Engine Applications," in IEEE Journal of the Electron Devices Society,

vol. 10, pp. 637-641, 2022, doi: 10.1109/JEDS.2022.3195119

[8] De S, Baig MA, Qiu B-H, Müller F, Le H-H, Lederer M, Kämpfe T, Ali T, Sung P-J, Su C-J, Lee Y-J and Lu DD (2022) Random and Systematic Variation in Nanoscale $Hf_{0.5}Zr_{0.5}O_2$ Ferroelectric FinFETs: Physical Origin and Neuromorphic Circuit Implications. *Front. Nanotechnol.* 3:826232. doi: 10.3389/fnano.2021.826232.

[9] S. De, D. D. Lu, H.-H. Le, S. Mazumder, Y.-J. Lee, W.-C. Tseng, B.-H. Qiu, Md. A. Baig, P.-J. Sung, C.-J. Su, C.-T. Wu, W.-F. Wu, W.-K Yeh, Y.-H. Wang, "Ultra-low power robust 3bit/cell $Hf_{0.5}Zr_{0.5}O_2$ ferroelectric finFET with high endurance for advanced computing-in-memory technology," in *Proc. Symp. VLSI Technology,* 2021.

[10] S. De, H.H.Le, B.H.Qiu, M.A.Baig, P.J.Sung, C.J.Su, Y.J.Lee and D.D.Lu, "Robust Binary Neural Network Operation From 233 K to 398 K via Gate Stack and Bias Optimization of Ferroelectric FinFET Synapses," in *IEEE Electron Device Letters*, vol. 42, no. 8, pp. 1144- 1147, Aug. 2021, doi: https://doi.org/10.1109/LED.2021.3089621.

[11] S. De, M. A. Baig, B. -H. Qiu, H. -H. Le, Y. -J. Lee and D. Lu, "Neuromorphic Computing with Fe-FinFETs in the Presence of Variation," *2022 International Symposium on VLSI Technology, Systems and Applications (VLSI-TSA)*, 2022, pp. 1-2, doi: 10.1109/VLSI-TSA54299.2022.9771015.

[12] De, S., Qiu, B.-H., Bu, W.-X., Baig, Md. A., Su, C.-J., Lee, Y.-J., & Lu, D. D. (2021). Neuromorphic Computing with Deeply Scaled Ferroelectric FinFET in Presence of Process Variation, Device Aging and Flicker Noise. *CoRR, abs/2103.13302.* https://arxiv.org/abs/2103.13302

[13] C. Matsui, K. Toprasertpong, S. Takagi and K. Takeuchi, "Energy-Efficient Reliable HZO FeFET Computation-in-Memory with Local Multiply & Global Accumulate Array for Source-Follower & Charge- Sharing Voltage Sensing," *2021 Symposium on VLSI Circuits*, 2021, pp. 1-2, doi: 10.23919/VLSI-Circuits52068.2021.9492448.

[14] S. Dutta *et al.*, "Monolithic 3D Integration of High Endurance Multi-Bit Ferroelectric FET for Accelerating Compute-In-Memory," *2020 IEEE International Electron Devices Meeting (IEDM)*, 2020, pp. 36.4.1-36.4.4, doi: 10.1109/IEDM13553.2020.9371974.

[15] Lehninger, D., Ellinger, M., Ali, T., Li, S., Mertens, K., Lederer, M., Olivio, R., Kämpfe, T., Hanisch, N., Biedermann, K., Rudolph, M., Brackmann, V., Sanctis, S., Jank, M. P. M., Seidel, K., A Fully Integrated Ferroelectric Thin-Film-Transistor – Influence of Device Scaling on Threshold Voltage Compensation in Displays. *Adv. Electron. Mater.* 2021, 7, 2100082. https://doi.org/10.1002/aelm.202100082

[16] F. Mo *et al.*, "Low-Voltage Operating Ferroelectric FET with Ultrathin IGZO Channel for High-Density Memory Application," in *IEEE Journal of the Electron Devices Society*, vol. 8, pp. 717-723, 2020, doi: 10.1109/JEDS.2020.3008789.

[17] C. Sun *et al.*, "First Demonstration of BEOL-Compatible Ferroelectric TCAM Featuring a-IGZO Fe-TFTs with Large Memory Window of 2.9 V, Scaled Channel Length of 40

nm, and High Endurance of 108 Cycles," *2021 Symposium on VLSI Technology*, 2021, pp. 1-2.

[18] Lu, D. D., De, S., Baig, M. A., Qiu, B.-H., and Lee, Y.-J. (2020). Computationally Efficient Compact Model for Ferroelectric Field-Effect Transistors to Simulate the On-line Training of Neural Networks. *Semicond. Sci. Technol.* 35 (9), 95007. doi: https://doi.org/10.1088/1361-6641/ab9bed.

Sourav De (Member, IEEE) works as a scientist at Fraunhofer IPMS, Center Nanoelectronic Technologies. He received his Ph.D. degree in Electrical Engineering in 2021 from National Cheng Kung University, and his B.Tech degree in Electronics and Communication Engineering in 2013, both from STCET, Kolkata. During his doctoral studies, Sourav De used to work in Taiwan Semiconductor Research Institute as graduate research student. Sourav worked towards integration of ferroelectric memory with advanced technology nodes. Sourav De joined the Fraunhofer Society in 2021. His main research interests are CMOS compatible emerging non-volatile memories for neuromorphic computing, advanced transistor and thin film transistor design for logic and memory applications, analog in-memory computing devices & circuits in CMOS and SOI technologies.

Sunanda Thunder works as a digital design engineering at TSMC, Taiwan. She received her Masters' degree from the department of International College of Semiconductor Technology from National Yang-Ming Chiao Tung University in 2021. She worked in Fraunhofer IPMS as a research intern before joining TSMC. Her primary research interests are neuromorphic circuit design with emerging non-volatile memories.

David Lehninger works as a project manager at Fraunhofer IPMS - Center Nanoelectronic Technologies. His main scientific interest is the optimization and integration of ferroelectric HfO2 films into the back end of the line of established CMOS technologies as well as the structural and electrical characterization of test structures and device concepts in the field of emerging non-volatile memories. Before joining Fraunhofer in 2018, he did a Ph.D. in Nanoscience at the TU Bergakademie Freiberg and a Master of Science in electrical engineering at the Dresden University of Technology.

Michael P.M. Jank received the diploma degree in electrical engineering from the Friedrich-Alexander-University of Erlangen-Nuremberg (FAU) in 1996, and the Dr.-Ing. degree with a thesis on extremely simplified routes to silicon CMOS devices in 2006. He started his career as a teaching assistant at the Chair of Electron Devices at FAU. Following the Dr.-Ing. degree in 2006, he joined the Fraunhofer Institute for Integrated Systems and Device Technology IISB, Erlangen, where he is currently heading the thin-film systems group, a joint undertaking with FAU. The group focuses on large area and printable thin-film electronics and develops materials, processing techniques, and thin-film devices based on of conventional PVD/CVD techniques as well as novel solution based approaches. He holds lectureships for Nanoelectronics and Printed Electronics from the FAU. He is reviewer for reknown international journals and contributes to scientific and industrial working groups on semiconductor memory devices and nanomaterials.

Maximilian Lederer is currently working as a project manager at the Fraunhofer IPMS Center Nanoelectronic Technologies. Prior, he finished his Ph.D. degree in Physics performing research in the field of ferroelectric hafnium oxide together with TU Dresden and Fraunhofer IPMS. He received his master degree in material science and engineering in 2018 at the Friedrich-Alexander Universität Erlangen-Nürnberg, Germany, and conducted a research semester at the Nagoya Institute of Technology, Japan, in 2017. His current research topics include non-volatile memories, neuromorphic devices, materials for quantum computing, structural and electrical analysis techniques as well as ferroelectrics.

Yannick Raffel was born in Dormagen, Germany, in 1993. He received the B.S. degree from the Department of physiks, Ruhr-Universität Bochum, Germany, in 2016 and the M.S. degree from the Department of physics (AFP), Ruhr-Universität Bochum, Germany, in 2018. In 2020 he is working toward the Ph.D. degree at the Fraunhofer-Institut CNT in Dresden, Germany. His current research interest is the investigation and description of low frequency noise and defect influences in nanotechnology.

Konrad Seidel received the Diploma degree in electrical engineering from the Dresden University of Technology, Dresden, Germany, in 2003. From 2004 to 2008, he was with the Reliability and Qualification Team of Flash Product Engineering, Infineon Technologies AG, Dresden, and Qimonda, Munich, Germany. Since 2008, he has been a Research Associate with Fraunhofer Center Nanoelectronic Technologies, Dresden, which is currently a Business Unit of the Fraunhofer Institute for Photonic Microsystems, Fraunhofer IPMS, Dresden. His current research interests include electrical characterization and reliability of integrated circuits as well as the integration and design of integrated high-density capacitors.

Thomas Kämpfe works as a senior scientist at Fraunhofer IPMS, Center Nanoelectronic Technologies. He received his Ph.D. degree in Physics in 2016 and his Diplom degree in Physics in 2011, both from TU Dresden, respectively. After research visiting scholar positions with the University of Colorado at Boulder and Stanford University, he joined the Fraunhofer Society in 2017. To date, Dr. Kämpfe authored and co-authored more than 150 peer-reviewed journal papers and conference proceedings. His main research interests are CMOS compatible ferroelectrics for advanced emerging memories, analog in-memory computing paradigms/architectures, high-frequency electronics, pyro- \& piezo-electronics as well as RF/mmWave devices & circuits in CMOS and SOI technologies.

Generating Trust in Hardware through Physical Inspection

Bernhard Lippmann, Matthias Ludwig, and Horst Gieser

Abstract—A globally distributed semiconductor supply chain, if not properly secured, may lead to counterfeiting, malicious modification, and IP piracy. Countermeasures like a secured design flow, including development tools, and physical and functional verification methods are paramount for building a trusted supply chain. In this context, we present selected image processing methods for physical inspection that aim to provide trust in the hardware produced. We consider aspects of the manufacturing process as well as the level of the physical layout. In addition, from the perspective of trust, we discuss the potential and risks of artificial intelligence compared to traditional image processing methods. We showcase the presented methods for a 28 nm process, and propose a quantitative trust evaluation scheme on the basis of feature similarities. A process for physical analysis and inspection that is free of anomalies or failures, of course, cannot be reached. However, from the trust point of view, it is crucial to be able to identify the sources of anomalies and clearly distinguish variations of the manufacturing process and artefacts of the analysis process from the signatures of potentially malicious activities. Finally, we propose a novel quantitative trust evaluation scheme which is partially based on the physical inspection methods outlined in this work.

Index Terms—hardware trust, reverse engineering, physical layout, manufacturing technology, physical verification, artificial intelligence, image processing

I. INTRODUCTION

SOCIETY depends increasingly on the availability, reliability and trustability of modern integrated circuits manufactured in nanoscale technologies.

These cover applications ranging from consumer, smart home, and internet of things (IoT) applications to autonomous driving and critical infrastructure. While being a key enabler for global megatrends like digitalization and decarbonization, trust in microelectronics is no longer granted by default. Globally distributed supply chains, outsourced semiconductor manufacturing, and the increased design and technology complexity extend the threat surface. These threats include the design of hardware Trojans, undocumented or optional functionality, access paths for programming and testing, counterfeiting, but also bugs or weak design solutions. Consequently,

This work was partly funded by the projects AI4DI, VE-FIDES, and platform project Velektronik. AI4DI receives funding within the Electronic Components and Systems for European Leadership Joint Undertaking (ECSEL JU) in collaboration with the European Union's Horizon2020 Framework Programme and National Authorities, under grant agreement no. 826060. VE-FIDES and Velektronik receive funding by the German Federal Ministry of Education and Research (grant no. 16ME0257 and grant no. 16ME0217).

(Corresponding author: B. Lippmann). B. Lippmann is with Infineon Technologies AG, Munich, Germany (e-mail: Bernhard.Lippmann@infineon.com). M. Ludwig is with Infineon Technologies AG, Munich, Germany (e-mail: Matthias.Ludwig@infineon.com). H. Gieser is with Fraunhofer EMFT, Munich, Germany (e-mail: Horst.Gieser@emft.fraunhofer.de).

these unwanted properties limit the trust in hardware solutions, particularly for critical applications. In addition, as most of the products are a combination of hard- and software, any hardware vulnerability must be treated as a complete system vulnerability independent of using a *secured* software solution and post-production patches are generally not available.

As hardware trust is no precise technical term with established measurement metrics, in a first approach, hardware trust is generated through security and functional testing as specified in various schemes like Common Criteria (CC) [1], FIPS 140 including the Cryptographic Module Validation Program (CMVP) [2], Trusted Computing Group Certification (TCG Certification) [3], the Security Evaluation Standard for IoT Platforms (SESIP) [4] and the Platform Security Architecture (PSA) [5]. They primarily cover on-device threats, including manipulative, side-channel, fault, or logical attacks. Both, the SAE Aerospace Standard (AS 6171B) and the IDEA 1010B Standard define inspection and test procedures for the identification of the suspect and counterfeit devices and target to verify a specific lot of devices from a nontrusted source [6], [7].

As previously elaborated, the complete microelectronic supply chain requires *trust* schemes. While the abovementioned methods provide generic, theoretical measures, these might not be comprehensive. The assessment of pre-silicon threats remains non-granularly resolved with a first approach shown in [8], where a metric considering the functional and structural test coverage has been introduced.

Various attack models and potential countermeasures have been discussed in scientific publications. These include split manufacturing [9] to tackle IP infringement or overproduction at

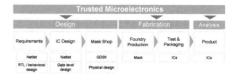

Fig. 1. Different stages of the IC development flow from the initial design phase until the final product.

untrusted manufacturing sites, layout integrity checking [10] via hardware reverse engineering to detect layout-bound hardware Trojans, cell camouflaging [11] to hamper reverse engineering for counterfeiting or the planning of subsequent hardware attacks, or logic locking [12] or finite state machine obfuscation [13] to protect against reverse engineering. Each of these methods can be assigned to one of the steps in the semiconductor supply chain shown in Fig. 1. While pre-production methods are mostly the preferred way, post-production, without further measures, analysis techniques are the only viable option. The second aspect to be motivated origins from the aforementioned system trust which is illustrated in Fig. 2.

In Fig.2, the different abstraction levels for computing or microelectronic systems are shown. Weaknesses or vulnerabilities in lower abstraction layers often

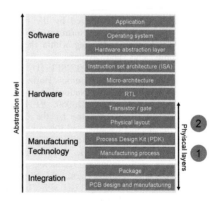

Fig. 2. Abstraction levels of computing systems. This work focuses on the physical layers of the abstraction stack with an emphasis on the physical layout and manufacturing technology.

lead to issues at higher abstraction levels, albeit deemed trustworthy or sufficiently secure. In these systems, the hardware acts as the root of trust. Bottom line, securing the lower system levels is of utmost importance. This work discusses two concrete ways of validating two abstraction levels. First ①, a method for validation of the manufacturing technology is elaborated. Second ②, methods for physical layout verification through hardware reverse engineering are discussed. These methods are distinguished into metal track segmentation, VIA detection, and standard cell analysis, which are further discussed from an algorithmic point of view. Finally, example analyses will be shown, and the viability of artificial intelligence (AI)-based methods will be discussed.

The contribution is organised as follows: Related work and the background for the verification processes of the physical layers and the major technical challenges are presented in Section II. Section III contains a selection of innovative methods for physical inspection. This includes a rule and AI-based image processing for layout and manufacturing technology assurance. The evaluation of these methods on a 28 nm test chip sample is demonstrated in Section IV. In section V, the trust evaluation scheme is elaborated.

II. BACKGROUND

Fig. 3 shows the major stages of a reverse engineering (RE) process used in academia, research institutes, commercial service providers, and integrated device manufacturers (e.g. [14], [15], [16], [17], [18], or [19]). The reverse engineering flow is constituted of a physical phase and a functional or electrical recovery phase. The flow is illustrated in Fig. 3 and explained in the following.

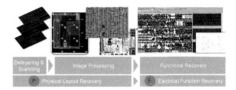

Fig. 3. Reverse engineering process overview.

The initial stage in the physical analysis lab is the preparation of samples. Each deposited physical layer of an integrated circuit is deprocessed and subsequently high resolution images of these layers are acquired using a scanning electron microscope (SEM). The complete scanned image mosaic is built upon several thousand individual tiles depending on the chip size. In the following step, we need to seamlessly stitch the individual tiles while analysing the shared overlapping area between these tiles. Geometrically undistorted mosaics of each layer are used for a correct 3D alignment of the complete scanned layer set [20]. During the layout recovery process, the images with wires and interconnecting structures are converted into a vector format. For identification and read back of digital or analogue devices from the raw layout images custom methods with domain expert's knowledge are required to solve these. As a result, we obtain a reconstructed device library (std. cell library) and the extracted connectivity between the devices. Via a back-annotation of individual devices, we generate a flat netlist of the analysed devices. Netlist interpretation algorithms are used to create an understanding of the extracted design, which can finally be verified through simulations and further electrical analysis.

Increased design complexity and shrinking technology nodes require reliable and innovative methods for the physical verification process. A compilation of these challenges is summarised

TABLE I
CHALLENGES FOR IC REVERSE ENGINEERING [21].

Task	Challenges
1. **Physical** layout recovery Ⓟ	• **Delayering (P1)** maximal uniform layer removal over complete chip size • **Technology (P2)** enable delayering of advanced nodes with ultra-thin and fragile inter-oxide layers, support Al & Cu technology, FinFET • **Chip Scanning (P3)** homogeneous, fast and accurate high resolution imaging over the complete chip area for all layers with minimal placement error • **Image Processing (P4)** precise layout recovery including preparation errors, indication of the error rate
2. **Electrical** function recovery Ⓔ	• **Digital Circuits (E1)** recovery and sense making of large digital circuits based on std. cell designs • **Analogue Circuits (E2)** recovery of circuit functions based on analogue devices with unclear electrical behaviour • **Robustness (E3)** robust to remaining errors coming from the physical layout extraction process
3. Analysis and scoring of **Security** protection mechanisms Ⓢ	• **Chip Individual Features (S1)** hardware security may include chip individual features like physical unclonable functions (PUFs), dedicated protection layers and protection circuits configured with run time keys, logic locking, etc. • **Design for Physical Analysis Protection (S2)** camouflaged cell libraries, timing camouflage • **Error and Effort Estimation (S3)** reliable indications must be shown how strong these measures are under the current analysis options

in Tab. I. A trustworthy verification process needs to address and solve these challenges with methods which are on itself trustworthy and reliable, while their limitations and potential failure modes must be explainable.

III. METHODOLOGY

Physical verification is defined by comparing physically measured data against a reference. The decision of authenticity (see section V) is based on these results. The first aspect of physical inspection described is the assurance provided by manufacturing technology, and the second aspect involves the inspection of the physical layout.

A. Manufacturing technology assurance

The technology can be subdivided into an assessment of the process design kit (PDK) on the one hand and the manufacturing process on the other hand. PDK aspects include physical aspects like

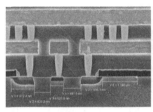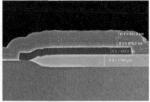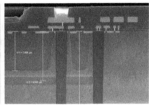

Fig. 4. Examples of labelled data showcasing the different ROIs: green – VIA; yellow – metal; teal – Local silicon Oxidation; red – poly; blue – deep trench isolation [22].

standard cell dimensions, static access memory cell dimensions, digital, analogue, and passive primitives, and design rules. The underlying manufacturing process covers e.g. physical aspects of different manufactured layers, critical dimensions, and utilised process materials. In this paper, we focus on the manufacturing process, although it cannot be sharply separated. A semiconductor device manufacturing process is a complex process, which is constituted of several hundred up to more than thousand individual sub-processes. The repeated sub-processes include, e.g. lithography, ion implantation, chemical-mechanical polishing, wet and dry etching, deposition, and cleaning. Three examples of how the process manifests itself in silicon are illustrated in Fig. 4. The images have been acquired after manual cross-sectioning of the respective semiconductor devices – post-production. The substantial different objects are shown colour-coded, and they describe following functionality (taken from [22]):

- **Metal:** Low resistance metallic connections between devices. Several metallisation layers can be stacked over each other to route inter-device connections.
- **Vertical interconnect access (VIA) / contact:** Low ohmic interconnections between different metallisation layers (VIA) or between devices and the lowest metallisation layer.

- **Lateral, shallow isolation (e.g. shallow trench isolation (STI) or local oxidation of silicon (LOCOS)):** Electrical lateral isolation between devices with a dioxide trough a *shallow* deposition.
- **Deep trench isolation:** Trenches for lateral isolation with a high depth-width ratio. Mostly found in analogue integrated circuits.
- **Polysilicon:** Poly-crystalline silicon which is used as gate electrode.

Besides a classification of the different classes, their respective features are of importance. These include geometrical ones (width, height, pitch) or material related ones. An example of how to measure properties of VIA1 and Metal3 is shown in Fig. 5. They may depend on the actual position of the cross section, the perspective and the SEM-parameters and require a careful calibration and awareness of uncertainty and process variations. Based on these features, the following hypothesis can be formulated:

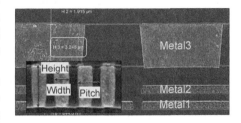

Fig. 5. Example cross-section image with annotated metal and contact/VIA features [22].

the entirety of a process can be interpreted as an individual manufacturing technology *fingerprint*. This hypothesis leads to the assertion that devices can be distinguished through their technological parameters. Consequently, device authenticity can be validated via a testing of aforementioned technological parameters against the expected parameters. These expected parameters are either provided by the manufacturer or extracted from known non-counterfeit devices, so called exemplars (SAE AS6171). The post-production technique allows the identification of counterfeit processes. The method covers all types of electronic devices (digital, analogue, systems-on-chip, FPGAs, etc.) and especially cloned, remarked, and repackaged types of counterfeits. For methodological details, refer to [22].

B. Physical layout verification

The verification of layout integrity can be executed through a comparison of reference layout data against the physical layout extracted in the recovery process. Layout recovery includes sample preparation and fast and accurate chip imaging, which needs to be addressed by a combination of the scanning electron microscope (SEM) hardware, including the detectors and the subsequent image processing algorithm. In this work, imaging is done using the RAITH 150Two chip scanning tool, which was also used in previous research [15], [18]. This work focuses on image processing as the last stage of the physical part of reverse engineering, i.e., the physical layout verification. All previous errors are manifested in this stage while error correction remains possible – if the entire process is well understood. Besides the final stage of the physical part, the imaging outputs are used for ensuing operations based on netlist interpretation and simulation. Consequently, the imaging output is a

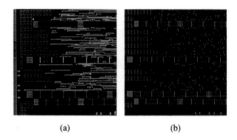

(a)   (b)

Fig. 6. Typical SEM images using two different detectors. (a) shows the scan with an InLens detector with a field of view of $30\,\mu m$ and a working distance of $10\,mm$. (b) shows the scan with ET-SE2 detector with a field of view of $20\,\mu m$ and a working distance of $10\,mm$.

suitable common thread for trust generation by physical inspection on the layout level. In this section, methods on different parts of the layout extraction process are discussed and made transparent.

Fig. 6 compares images generated using an Everhart-Thornley-SE2 (ET-SE2) detector and an InLens detector available in the RAITH 150Two. As the material contrast is increased with the ET-SE2 detector, large and small layout structures become bright, and the background remains dark, enabling a threshold-based segmentation approach. Using the InLens detector, the brightness of a metal line depends on its structural size. Large structures in the left area of the reference image appear darker in their centre area. Solving these challenges is vital for a successful extraction.

1) Metal track segmentation: A threshold-based extraction of metal lines using algorithms like Otsu, Li ([18], [15]) is performed and confirmed within our analysis projects (Fig. 7) on images with sufficient separation between fore- and background colour levels.

The challenge arises when the colour of thick metal patterns drops close to the background colour value, as shown in Fig. 8. This is solved with a customised segmentation algorithm.

The presented algorithm, called *SEM-Seg* (see Fig. 8 and 9), is based on

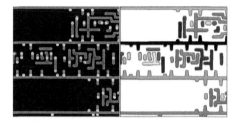

Fig. 7. Image segmentation using threshold algorithm.

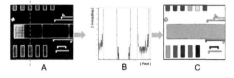

Fig. 8. Image segmentation using custom algorithm *SEMSeg*.

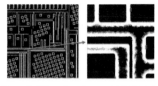

(a) Stage 1: Edge direction analysis (red: identified background, blue: max edge colour value, yellow: foreground).

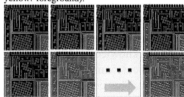

(b) Stage 2: Iterative colouring identified background.

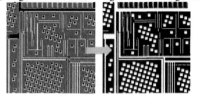

(c) Stage 3: Flip colour levels.

Fig. 9. *SEMSeg* algorithm, segmentation of SEM images with overlapping foreground and background colour level.

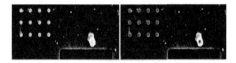

Fig. 10. Contact detection using Hough algorithm.

the identification of fore- or background area depending on the gradient of the edge shape rather than relying on absolute colour value. In our initial stage, we inspect these edge points and identify the factual background on one side of the edge. In stage 2 of our algorithm, we sequentially convert identified background area into a red-coloured area using a flood-fill algorithm until only white edges remain or black not yet flipped foreground inside larger structures remain. Finally, we need to convert the colour values to the segmentation target values, the foreground becomes white, and the background is changed to black. Fig. 9 shows interim stages of the segmentation flow, leading to an error free-segmentation, where threshold-based algorithms would be infeasible.

2) VIA segmentation with customised Hough algorithm: The identification of contacts located on the top side of a depassivated metal layer can be achieved using the Hough circle transformation (HCT) [23] with the risk of false contact detection using no optimised parameters or due to the limitations of the HCT.

As shown in the right image of Fig. 10, one wrong contact has been found on a particle.

In the original HCT each pixel is evaluated concerning its possibility of being located on a circle with a radius r. This result is stored in the accumulator matrix. In the final step, selecting local maxima from the accumulator matrix returns the possible circles. We introduce a rule-based modified accumulator calculation and evaluation considering, e.g., a colour change over the contact area or an evaluation of the outer circle area aiming to remove false contacts maxima from the accumulator. The accumulator result in Fig. 11 shows the dominant peak for the particle, which is eliminated and finally generates an error-free contact identification.

3) Standard cell analysis:

a) Automatic standard cell identification: The automatic identification of logic gates in a given design is based on domain knowledge about design principles and customised image processing algorithms. The complete process contains three major steps as shown Fig. 12.

The major innovation uses the fact that our standard cell images display polysilicon, contacts and p/n-doped area at the same time (left side of Fig. 13). As these structures are brighter than the background (isolation), in a black-white-image of the complete cell array a flood-fill mechanism, stopping only at the isolation area, allows the identification of the correct cell dimension perpendicular to the power line direction. By this procedure, the complete image is automatically segmented into the individual std. cells (right side of Fig. 13).

b) Standard cell structure analysis: We continue the analysis of std. cell images with a customized image processing algorithm converting the identified cells in a flat transistor level netlist. First, we segment the contact, polysilicon and active area from an std. cell image. The active area is not completely visible in the input image, so it is filled by connecting the paths between two active area segments, which are covered

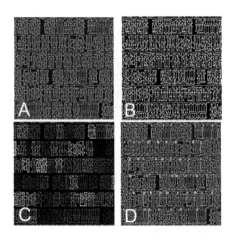

Fig. 12.   Standard cell identification process. (a) SEM image displaying poly-silicon, active area, and contacts (b) Detection of power lines (V_{DD}, V_{SS}). (c) Segmented standard cells using custom image processing algorithms. (d) Classification of different standard cells.

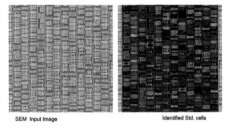

Fig. 13.   Segmentation of polysilicon, active area delayering SEM images into standard cells using a custom algorithm with domain knowledge.

by polysilicon. We construct nothing else like the channel area of the transistor in this step (B). The logical combination between the completed active area and the polysilicon shapes on the pixel base defines the active transistor area. Segmenting this image, we obtain the active transistor area as polygons (C, D). The transistor definition algorithm now extends the active transistor area until half of the contact distance, and finally, we build transistors and have the net information adding the M1 layer. The extracted geometrical shapes are stored

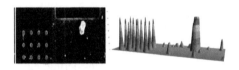

Fig. 11.   Contact detection using modified Hough algorithm.

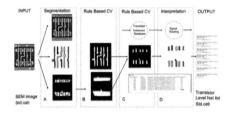

Fig. 14.   Extracting transistor level netlist from std. cell SEM images displaying polysilicon, contact, and active area layouts.

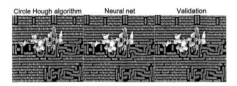

Fig. 15.   Three different VIA extraction methods [17].

in a vector format, and we use a layout versus schematic-like back annotation tool for the generation of an electrical transistor level netlist.

4) Discussion on the applicability of deep learning approaches: For a more robust and better-yielding identification of VIAs, a convolutional neural network (CNN) based architecture is used. A training set is obtained from images generated through classical processing and following manual cleaning. The left image in Fig. 15 shows the detection of VIAs using the customized Hough transformation. Results obtained from a pure deep learning approach are shown in the center image. A hybrid approach using classical Hough transformation for detection and deep learning (DL) based contact validation for improved computing performance is shown in the right image.

IV. EVALUATION OF PHYSICAL LAYOUT EXTRACTION

In our latest work, the developed physical and functional reverse engineering workflow has been applied on test chips manufactured at the 28 nm technology node.

A. Sample preparation

The delayering process has been continuously improved presenting homogeneous sample preparation techniques almost over the entire chip area at the 40 nm technology node [21]. Due to the ever-shrinking inter-layer oxide thicknesses for the 28 nm node a homogeneous full chip area delayering for the lower metal levels and the polysilicon level could not be achieved as shown in Fig. 17. However, dedicated larger circuit blocks are available in adequate delayering quality.

The delayering quality assessment is based on a first inspection of the colour-uniformity of the whole chip module with optical microscopy techniques. Colour changes inside one layer indicate different modules and different remaining layers stacks. Finally, SEM inspection

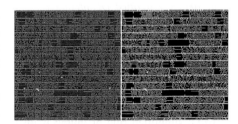

Fig. 16.   Metal layer extraction with deep learning [17].

Fig. 17.   28 nm technology, delayering study on test samples. Optical microscope and SEM images are used for quality assurance.

allows the precise and detailed quality assessment.

B. Layout recovery

Various image processing methods including the discussed solutions have also been successfully applied to layout extraction studies using 28 nm test chip samples.

Fig. 18 shows high-resolution sample images and the corresponding extracted layout patterns.

Successful VIA recovery is demonstrated in Fig. 18a and 18b. Even new challenges arise from large VIA shapes covering the metal pattern underlying. We implemented dedicated image preprocessing algorithms using the design rules (DR) for recovering the hidden or over-blended metal pattern. Fig. 18d displays a recovered part of a 28 nm layout. Fig. 18c shows the successful polygon recovery even due to scan time contains the scanning resolution has been reduced.

C. Comparison of layout recovery using classical CV against DL methods

A comparative study between layout recovery using classical image processing algorithms and DL-based solutions has been executed. Obtained results show a dependence on image quality and applied training efforts and training data. An overall simplified rule of thumb is given in Tab. II. We observed a performance increase with the drawbacks of training data dependency, sample-specific training, and labelling efforts.

(a) VIA detection: Thick contact pattern, partial coverage of metal lines below.

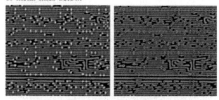

(b) VIA detection: Polygon shaped VIAs.

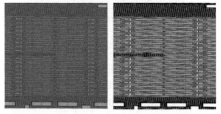

(c) Metal line extraction: larger structures, regular layout.

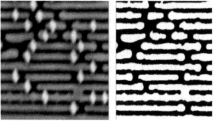

(d) Metal lines extraction: detail view of small and non-regular metal tracks, min. pitch between wires.

Fig. 18. Overview of different metal segmentation and VIA detection tasks of the 28 nm test chip sample.

Images obtained from smaller technology nodes with a lower image quality achieved better conversions results using AI methods. Classical image processing methods show a much higher yield drop. During the quality assessment of image sets using DL-based image processing we observed unexpected failures as shown in Fig. 19 as only a minor brightness

TABLE II
RULE OF THUMB PERFORMANCE OPTIONS FOR
THE EXTRACTION OF CLASSICAL VS. DL-BASED
VIA DETECTION.

	Classical	Deep Learning	Remark
Good image quality	99.5%	99.6%...100%	VIAs perfectly visible
Poor image quality	∼ 50%	∼ 90%	VIAs hard to recognise

Fig. 19. Polygon extraction example with deep learning.

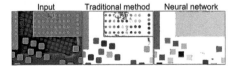

Fig. 20. Reconstruction capabilities DL.

change over the entire image area led to significant conversion failures.

V. TRUST ASSESSMENT THROUGH PHYSICAL INSPECTION

A. Introduction of similarity metrics

HG: Erst hier kommt das Package ins Spiel, daher ist ist daer andere Counterfeit claim zu früh.

The complete physical verification process is composed of individual comparisons of measurable product features against the original data covering the discussed aspects of the manufacturing technology and the reconstruction of the geometrical layout. Furthermore, albeit not elaborated in this work, the respective physical design kit (PDK) and physical die package information are two more pillars for a trust validation flow. This flow is described in the following mathematically.

$$S(M,G)$$
$$= \frac{1}{N}\big(PackCom(M,G)$$
$$+ DFCom(M,G)$$
$$+ TFCom(M,G)$$
$$+ PolyCom(M,G)\big)$$

$$= \frac{1}{N}\bigg(\sum_{Package}^{N_1} PackCom_i(M_i,G_i)$$
$$+ \sum_{PDK}^{N_2} DFCom_i(M_i,G_i)$$
$$+ \sum_{Tech.}^{N_3} TFCom_i(M_i,G_i)$$
$$+ \sum_{Layout}^{N_4} PolyCom_i(M_i,G_i)\bigg) \quad (1)$$

In (1), $S(M,G)$ describes the similarity been a measured product M and the reference data G (i.e. golden model). Each comparator consists of a number N_i of single feature comparisons. A single stage comparator function should measure the similarity and returns 1 for a perfect match and 0 if no correlation to the reference data has been found. For thickness and dimension measurements in the technology comparator can be directly applied, polygon similarity measurements based on a polygon-based F1 score or a pixel-based XOR score have been presented in [17]. For normalising the complete similarity function $S(M,G)$ we need to add the factor $\frac{1}{N}$ with $N = \sum_i N_i$.

The proposed comparator definitions include:

- **Technology Feature Comparator (TFCom):** Compares wafer fabrication features like deposited layers including minimum structure width, thickness and material against the original technology.

  - **Efficiency:** very high, only SEM cross-section images needed
  - **Use case:** manufacturing technology fingerprinting

- **Package Feature Comparator (PackCom):** Analogue to the TFCom, but using die package manufacturing steps, including labelling processes with font size and marking details.

- **Efficiency:** typically used tools like optical microscopes, height gauges and x-ray systems offer high throughput at low analysis costs.
- **Use case:** inspection of bond wires, bond scheme, device marking including acetone wipe and scrub test, die marking can detect remarked and recycled devices. Confocal Scanning Acoustic Microscopy C-SAM may be used to detect delamination of reused devices [6].

• **Polygon Comparator (PolyCom):** After complete recovery of the physical design the comparison of the recovered polygons against the original layout can be executed. Challenges from Tab. I (Ⓟ, Ⓢ) apply as large area chip delayering, scanning and image analysis are required. Unstable and imperfect delayering processes may require in addition scanning dedicated to special features e.g. VIAs or the use of a larger sample-sets, both increasing costs and time.

- **Efficiency:** dedicated highly customised tools are necessary
- **Use case:** the most sensitive method for detection of circuit modifications presented here

• **Design Features Comparator (DFCom):** Besides the manufacturing technology, a set of logic standard cells, analogue primitive devices, SRAM, ROM, NVM cells and modules are included in the PDK. Checking the physical layout in dedicated local areas the recovery of many PDK features can be performed.

Using these data an extended manufacturing technology footprint is generated without the need for a complete layout recovery. As the polygon comparator is just based on the similarity of different geometrical layout descriptions and product specific, the *DFCom* is providing an electrical interpretation of the recovered patterns. Furthermore, in combination with the technology comparator, it defines a complete manufacturing technology fingerprinting.

- **Efficiency:** does not require the complete layout recovery, significantly higher compared to a full polygon comparison
- **Use case:** technology fingerprinting

B. Discussion on trust metrics

The trust generated by a comprehensive physical verification can therefore be modelled by assigning individual weights to each comparison stage as shown by (2). This extends previously defined $S(M, G)$ to $T(M, G)$. In this adapted equation, individual weights (w) allow an adaptive rating of different features but also within different comparators.

$$
\begin{aligned}
T(M,G) \\
= \frac{1}{N} \Bigg(& \sum_{Package}^{N_1} w_{Package}(i) \\
& \cdot PackCom_i(M_i, G_i) \\
+ & \sum_{PDK}^{N_2} w_{PDK}(i) \cdot DFCom_i(M_i, G_i) \\
+ & \sum_{Tech.}^{N_3} w_{Tech.}(i) \cdot TFCom_i(M_i, G_i) \\
+ & \sum_{Layout}^{N_4} w_{Layer}(i) \\
& \cdot PolyCom_i(M_i, G_i) \Bigg)
\end{aligned} \tag{2}
$$

As given in (3), in practice not all features are measured by physical inspection and since many comparator stages (Comparator(i,j)) are costly and time-consuming, normally, only a limited number of features is selected.

$$T(M,G)$$

$$= \frac{1}{N} (\underbrace{\sum_{j=0}^{CT} \sum_{i=0}^{N_j} \underbrace{w_j(i)}_{\text{\textcircled{W}}}}_{\text{test coverage}}$$

$$\cdot \underbrace{Comparator_{(i,j)}(M_i, G_i))}_{\text{\textcircled{C}}} \quad (3)$$

Under these constraints, a realistic trust evaluation with skipped comparator stages must still provide a trustworthy test coverage. CT in (3) is the number of used comparator types. As the assigned weights \textcircled{W} to individual stages express their impact on the overall hardware stages with an assumed low weight might be skipped.

Still, this quite intuitive approach yields two major challenges: First, as many input data for \textcircled{C} are a result of quite challenging lab processes, we have to consider a concise handling of the variations and tolerances each manufacturing and analysis process as well as the metrology has. Furthermore, there may be published documented and undocumented changes in the process of the original device manufacturer. The rule-based image processing algorithms and results are presented above but no systematic approach for analysis of method failures is existing. This holds specifically true for ML/DL-based comparators where training and analysis data play an important role.

Secondly, the computation of expressive values for the weights is far from trivial. A quantification of trust and security is one a big challenge as concluded in

the common reverse engineering scoring system (CRESS) [24]. Nonetheless, a full practical realisation exists for software in the form of the common vulnerability scoring system (CVSS) [25]. In the end, a comprehensive method for trust evaluation, as proposed in this work, is a dynamic one. Ultimately, the decision how to scale the weights and which methods to use differs from use case to use case and a myriad of factors (e.g. end application, economic factors, potential safety effects, etc.) must be taken into account.

VI. CONCLUSION

Besides a secured design flow, covering electronic design automation tools and methodology and a secured manufacturing process, physical inspection is one pillar to create hardware trust. The implementation techniques to generate trust through physical inspection must be profoundly understood when applied. Furthermore, it must be acknowledged that physical constraints in sample preparation, imaging and scanning may lead to imperfect comparator inputs and must be carefully investigated. Continuously executed analysis projects will demonstrate the achieved trust level while aiming to reduce the test coverage.

Addressing the requirements for an advanced physical verification process, several innovative process solutions have been developed and evaluated. Novel image processing algorithms using dedicated expert semiconductor analysis knowledge allow the effective and precise extraction of layout and technological information. With the introduction of advanced technology nodes, AI-based image processing can provide additional options. Creative solutions are needed as the number of challenging imaging problems increases.

VII. FUTURE WORK

Considering trust through testing against well-defined criteria with accepted frameworks as a reference, physical verification lacks these generally accepted methods and verification flows. As AI-based image processing methods need to prove their trustworthiness and performance increase over conventional image processing, a continuation of benchmarking is mandatory. These will be the two major challenges for the future of effective and efficient physical inspection flows. Still, manufacturing technology, physical layout, and functional verification are major building blocks of a complete future product verification process. For future work, practical analysis projects will provide concrete numerical results and contribute to the overall trust level.

ACKNOWLEDGMENT

The authors would like to thank Anja Dübotzky and Peter Egger of Infineon Technologies AG for physical sample preparation and Tobias Zweifel and Nicola Kovač of Fraunhofer EMFT for EBeam scanning.

REFERENCES

[1] *The Common Criteria*, https : / / www . commoncriteriaportal . org/, Accessed: 2022-08-06.

[2] A. Vassilev, L. Feldman, and G. Witte, *Cryptographic Module Validation Program (CMVP)*, en, 2021-12-01 2021. https://tsapps.nist.gov/publication/get_pdf.cfm?pub_id=917620.

[3] *TCG Certification Programs*, en. https : / / trustedcomputinggroup . org /membership/certification/.

[4] *SESIP: An optimized security evaluation methodology, designed for IoT devices*, en. https://globalplatform.org/sesip/.

[5] *Platform Security Model version 1.1*, en, 2019-12-01 2019. https : //www.psacertified.org/app/uploads/2021/12/JSADEN014_PSA_Certified_SM_V1.1_BET0.pdf.

[6] U. Guin, D. DiMase, and M. M. Tehranipoor, "A Comprehensive Framework for Counterfeit Defect Coverage Analysis and Detection Assessment," *Journal of Electronic Testing*, vol. 30, 2014.

[7] *Test Methods Standard; General Requirements, Suspect/ Counterfeit, Electrical, Electronic, and Electromechanical Parts AS6171*, en, Oct. 2016. https : / / www . sae .org/standards/content/as6171/.

[8] J. Cruz, P. Mishra, and S. Bhunia, "INVITED: The Metric Matters: The Art of Measuring Trust in Electronics," in *2019 56th ACM/IEEE Design Automation Conference (DAC)*, 2019.

[9] M. Jagasivamani *et al.*, "Split-fabrication obfuscation: Metrics and techniques," in *2014 IEEE International Symposium on Hardware-Oriented Security and Trust (HOST)*, IEEE, 2014.

[10] M. Ludwig, A.-C. Bette, and B. Lippmann, "ViTaL: Verifying Trojan-Free Physical Layouts through Hardware Reverse Engineering," in *2021 IEEE Physical Assurance and Inspection of Electronics (PAINE)*, (Dec. 2021), preprint: http://dx.doi.org/10.36227/techrxiv.16967275, Washington, DC, USA: IEEE, Dec. 2021.

[11] A. Vijayakumar *et al.*, "Physical Design Obfuscation of Hardware: A Comprehensive Investigation of Device and Logic-Level Techniques," *Trans. Info. For. Sec.*, vol. 12, no. 1, Jan. 2017.

[12] S. M. Plaza and I. L. Markov, "Solving the Third-Shift Problem

in IC Piracy With Test-Aware Logic Locking," *IEEE Transactions on Computer-Aided Design of Integrated Circuits and Systems*, vol. 34, no. 6, Jun. 2015.

[13] M. Fyrbiak *et al.*, "On the Difficulty of FSM-based Hardware Obfuscation," *IACR Transactions on Cryptographic Hardware and Embedded Systems*, Aug. 2018.

[14] R. Torrance and D. James, "The state-of-the-art in semiconductor reverse engineering," in *48th ACM/EDAC/IEEE Des. Automat. Conf. (DAC)*, Jun. 2011.

[15] *The Role of Cloud Computing in a Modern Reverse Engineering Workflow at the 5nm Node and Beyond*, vol. ISTFA 2021: Conf. Proc. 47th Int. Symp. for Testing and Failure Anal. Oct. 2021.

[16] A. Kimura *et al.*, "A Decomposition Workflow for Integrated Circuit Verification and Validation," *J. Hardw. Syst. Secur.*, vol. 4, Mar. 2020.

[17] B. Lippmann *et al.*, "Verification of physical designs using an integrated reverse engineering flow for nanoscale technologies," *Integration*, vol. 71, Nov. 2019.

[18] R. Quijada *et al.*, "Large-Area Automated Layout Extraction Methodology for Full-IC Reverse Engineering," *Journal of Hardware and Systems Security*, vol. 2, 2018.

[19] H. P. Yao *et al.*, "Circuitry analyses by using high quality image acquisition and multi-layer image merge technique," in *Proceedings of the 12th International Symposium on the Physical and Failure Analysis of Integrated Circuits, 2005. IPFA 2005.*, Jun. 2005.

[20] A. Singla, B. Lippmann, and H. Graeb, "Recovery of 2D and 3D Layout Information through an Advanced Image Stitching Algorithm using Scanning Electron Microscope Images," Jan. 2021.

[21] B. Lippmann *et al.*, "Physical and Functional Reverse Engineering Challenges for Advanced Semiconductor Solutions," in *2022 Design, Automation & Test in Europe Conference & Exhibition (DATE)*, IEEE, Mar. 2022.

[22] D. Purice, M. Ludwig, and C. Lenz, "An End-to-End AI-based Automated Process for Semiconductor Device Parameter Extraction," in *Industrial Artificial Intelligence Technologies and Applications*. Vienna, Austria: River Publishers, 2022, ch. 4, pp. 53–72.

[23] U. Botero *et al.*, *Hardware Trust and Assurance through Reverse Engineering: A Tutorial and Outlook from Image Analysis and Machine Learning Perspectives*, Oct. 2020.

[24] M. Ludwig, A. Hepp, M. Brunner, and J. Baehr, "CRESS: Framework for Vulnerability Assessment of Attack Scenarios in Hardware Reverse Engineering," in *2021 IEEE Physical Assurance and Inspection of Electronics (PAINE)*, (Dec. 2021), preprint: https://dx.doi.org/10.36227/techrxiv.16964857, Washington, DC, USA: IEEE, Dec. 2021.

[25] FIRST.Org, Inc. (Aug. 21, 2022). "Common Vulnerability Scoring System version 3.1: Specification Document," https://www.first.org/cvss/specification-document.

Bernhard Lippmann received a diploma degree in Physics from the Technical University Munich (TUM), Germany in 1992. He started his career at Hitachi Semiconductor Europe in Landshut in the type engineering group. From 1993 until 1998 he was responsible for the physical and electrical failure analysis and yield enhancement programs for several generations of DRAM and smart card products. In 1999 he joined Infineon Technologies AG (Siemens) former Chipcard and Security Division in Munich. He is responsible for competitor analysis, reverse engineering and benchmarking projects. Currently, he is responsible for the project coordination of public-funded project RESEC (https://www.forschung-it-sicherh eit-kommunikationssysteme.de/projekte/resec) at the Connected Secure Systems (CSS) division of Infineon. He holds several patents on smart card and security topics, his publications cover Java Card benchmarking and circuit reverse engineering.

Horst A. Gieser is head of the AT (Analysis and Test) team at the Fraunhofer-Institution for Microsystems and Solid State Technologies EMFT (www.emft. fraun-hofer.de). He received his diploma in Electrical Engineering and his Ph.D. from the Technical University in Munich where he started his first laboratory and research team for analysis and test in 1989 and has transferred it to Fraunhofer in 1994. Starting and growing with Electrostatic Discharge ESD, he has extended his research and application interest into the field of the analysis for Trusted Electronics down to the nanoscale and the cryo-characterization of quantum devices. His lab is CC-EAL6 certified for the physical analysis of security chips. Mainly in the field of ESD he has authored and contributed to more than 120 publications including several invited talks at international conferences in the US, Taiwan and Japan. He is author of several publications in peer reviewed journals. Four publications won awards. Today he is leading activities in several public funded projects on Trusted Electronics.

Matthias Ludwig received the B.Eng. from Regensburg University of Applied Sciences, Germany in 2017, and the M.S. from Munich University of Applied Sciences, Germany in 2019, both in electrical and computer engineering. He is currently pursuing his Ph.D. with the department of electrical and computer engineering at Technical University of Munich, Germany and is with the Connected Secure Systems (CSS) division of Infineon Technologies AG, Neubiberg, Germany. His research interests include hardware security with a focus on anti-counterfeiting, hardware trust and physical security.

Meeting the Latency and Energy Constraints on Timing-critical Edge-AI Systems

Ivan Miro-Panades, Inna Kucher, Vincent Lorrain, and Alexandre Valentian

Abstract—Smart devices, with AI capabilities, at the edge have demonstrated impressive application results. The current trend in video/image analysis is to increase its resolution and classification accuracy. Moreover, computing object detection and classification tasks at the edge require both low latency and high-energy efficiency for these new devices. In this paper, we will explore a novel architectural approach to overcome such limitations by using the attention mechanism of the human brain. The latter allows humans to selectively analyze the scene allowing limiting the spent energy.

Index Terms—Edge AI accelerator, high-energy efficiency, low-latency, object detection.

I. INTRODUCTION

THE observed trend, in visual processing tasks, is to increase the complexity of neural network (NN) topologies to improve the classification accuracy. This results in NN models being deeper and larger leading

This work was in part funded by the ECSEL Joint Undertaking (JU) under grant agreement No 876925. The JU receives support from the European Union's Horizon 2020 research and innovation program and France, Belgium, Germany, Netherlands, Portugal, Spain, Switzerland

Authors Ivan Miro-Panades and Alexandre Valentian are with Univ. Grenoble Alpes, CEA, List, F-38000 Grenoble, France (e-mail: ivan.miro-panades@cea.fr and alexandre.valentian@cea.fr).

Authors Inna Kucher and Vincent Lorrain are with Université Paris-Saclay, CEA, List, F-91120, Palaiseau, France (e-mail: inna.kucher@cea.fr and vincent.lorrain@cea.fr)

to several issues when used in edge applications. Even though mobile versions of some network topologies have been introduced over time, it remains difficult to integrate them on-chip in an energy-efficient manner. The main issue is the large number of parameters, requiring the use of an external memory which leads to a large power dissipation due to the data movement. It should be noted that the energy necessary for moving data is three orders of magnitude larger than that spent for doing computations on the same data [1]. This is the primary issue ("Issue No. 1") that must be addressed.

Moreover, when processing a video input stream, the whole image is processed, frame by frame, even though there is enormous spatial redundancy between consecutive frames. If the target application requires a low reaction time to events (e.g., to an object or person moving), the frame rate needs to be high, leading to high instantaneous power values. On the other hand, if the frame rate can be kept low, such as for security surveillance applications, the system overall energy efficiency would still be poor, because of the high inter-frame redundancy (especially during nights and weekends). The second issue ("Issue No. 2") that must be addressed is to reconcile low power dissipation and short reaction times to events.

In those respects, bio-inspired approaches can lead to innovative solutions. For instance, neuroscientists have found anatomical evidence that there are two separate cortical pathways, or 'streams', in the visual cortex of monkeys [2]: the ventral pathway and the dorsal pathway, as shown in Fig. 1. On one hand, the dorsal stream is relatively fast, sensitive to high-temporal frequencies (i.e. motion), viewer-centered and relatively unconscious. It has been called the "Where" path, since it is used for quickly retrieving the location of objects, especially of moving ones. On the other hand, the ventral stream is relatively slow ($\sim$4x longer reaction time), sensitive to high-spatial frequencies (i.e., details), object-centered and relatively conscious. It is known as the "What" path, involved in the recognition of objects. If millions of years of evolution have led to such a 2-path solution, this is because it brought a competitive advantage to our ancestors, allowing them to quickly evade threats, even before their brain knew the nature of the threat.

Even though this 2-stream hypothesis has been disputed over the years and we now know that those pathways are not strictly independent, but do interact with each other (for instance for skillfully grasping objects [3]), it is still relevant to the problem at hand, as it provides a good fit to many motor and perceptual findings.

In this work, we focus on the "Where" subsystem, as the "What" one is already well addressed with existing accelerators [10]. The main objectives are therefore to obtain the lowest possible latency and power values. First, we have started by selecting an adequate neural network topology, i.e., one with a small number of parameters, but with only slightly degraded performances: we have chosen the MobileNet-V1 topology [4]. Since all the parameters need to be stored on-chip, for solving 'Issue N° 1', the synaptic weights and activation values must be heavily quantized, without a significant loss in accuracy our in-house learning framework was complemented with a state-of-the-art quantization-aware training (QAT) algorithm. This tool is now available in open source and presented in Section II. An innovative architecture was considered for the hardware, once again taking inspiration from biology: layers V1 to V3 of the visual cortex, which are sensitive to orientations of edges and to movement, are fixed early on during life. For instance, V1 undergoes synaptic and dendritic refinement to reach adult appearance at around 2 years of age [5]. Even though these synaptic weights will not be learnt again during adulthood, that does not prevent our visual cortex to learn how to recognize new objects. We have thus chosen to fix the feature extraction layers of the MobileNet once and for all (while ensuring they remain sufficiently generic) and then to apply a transfer learning technique, to target several applications. Fixing synaptic weights actually leads to tremendous energy and latency savings, e.g. getting rid of memory accesses. Such an architecture can be used in an attention mechanism, solving 'Issue N° 2'. The architecture analysis is described in Section III. Finally, Section IV concludes this work.

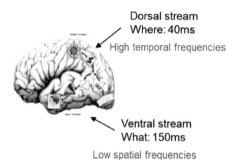

Dorsal stream
Where: 40ms
High temporal frequencies

Ventral stream
What: 150ms

Low spatial frequencies

Fig. 1. Illustration of the two visual pathways or streams in the visual cortex, used for extracting different information.

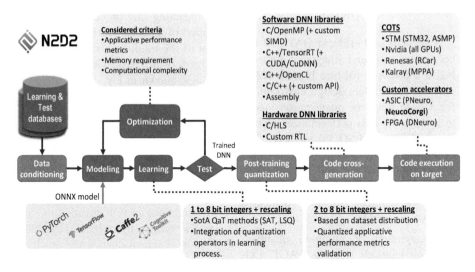

Fig. 2. N2D2 framework.

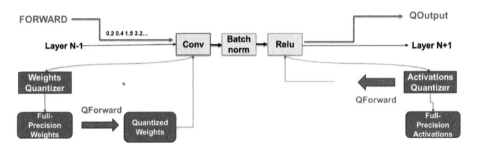

Fig. 3. Forward and backward quantization passes.

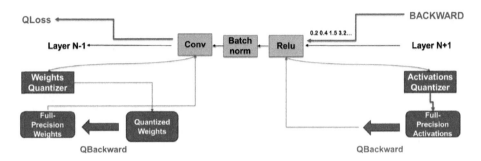

Fig. 4. Backward propagation with QAT.

II. QUANTIZATION AWARE TRAINING

A scalable and efficient QAT tool has been developed and integrated into the N2D2 framework [6] (see Fig. 2). N2D2 provides a complete design environment for a wide range of quantization modes to achieve the best performances including SAT [7] and LSQ [8] methods. The overall framework and the addition of the quantization aware training modules are shown in Figs. 3 and 4 above.

The advantages of this dedicated framework include:

- Integration of common quantization methods (Dorefa, PACT, CG-PACT).
- Straightforward support of mixed-precision operator (2-bits to 8-bits on Weights and/or Activations).
- Automatic support of non-quantized layers (e.g., batch normalization).
- Training phase based on optimized computing kernels, resulting in fast evaluation of the quantization performance.

There are two separate quantization modules, one dedicated to weights and another one to activations. This is illustrated in Figs. 3 and 4 above, where the example Layer N consists of a Convolutional layer followed by the Batch normalization layer with activation (typically ReLu). The weights of this Convolutional layer are quantized to a desired precision, using the quantize_wts function. Batch normalization stays in full precision and goes through the activation function. This output is then quantized to the required precision, using the quantize_acts function. It must be noted that the two quantification precisions, i.e., of the weights and activations, might not necessarily be the same.

During the neural network training, the parameters are adjusted using backprop-agation of errors on these parameters, and repeating this process for a certain number of epochs, until the figure of merit of the training is satisfactory.

The forward pass with QAT follows the logic shown in Fig. 3:

- Inputs arriving to the convolutional layer are passed through the convolution operation, where the weights are quantized beforehand using Weight Quantizer module;
- The output is propagated to Batch normalization layer, which provides operations in full precision;
- The output from Batch normalization is transformed to its quantized values using the Activation Quantizer;
- At the end, the quantized output is passed as an input to the next layer.

The backward propagation with QAT, shown in Fig. 4, includes the following steps:

- Starting from the errors on quantized activations, the errors on full precision activations are computed, using the derivatives of the transformations applied in the forward pass;
- Then these errors, on full precision activations, are propagated through Batch normalization and convolutional layers;
- In a similar way, the errors on full precision weights are computed using quantized weights errors.

During the learning procedure, both full precision and quantized quantities are kept. One has to keep in mind that, during the training, the applied procedure is called "fake" quantization, since even quantized values are kept using floating-point type.

Once the network is trained, the weights and inputs are transformed into true integer values before execution on a hardware target.

III. ARCHITECTURE EXPLORATION

The architecture exploration started with the choice of the NN topology, with a low energy and low latency per inference in mind: the target is an energy below 4mJ per image (HD: 1280x720 pixels) and a latency compatible with a 30FPS frame rate (i.e. below 30ms). A tradeoff must thus be made between network complexity and operations per inference. A lower number of operations obviously leads to a lower number of Multiplication-Accumulation (MAC) operations to be performed per image.

Fig. 5 illustrates the various topologies that can be found in the literature [9]. The MobileNet-V1 topology has been chosen, as it uses depth-wise and point-wise convolutions to reduce the computing complexity (their difference with a standard convolution is shown in Fig. 6).

Usually, NN accelerators use a layer-wise architecture. This makes it possible to support different topologies, since networks are computed layer-wise. It also makes it possible to compute multiple images per layer, i.e., inputs with a batch size higher than one: the synaptic weights are read once and can be reused for the different images, reducing the power dissipation. However, in our case, we have conflicting constraints: the topology is fixed and the batch size is equal to one to limit the processing latency. A streaming architecture is thus considered, since the latency is minimized and the fixed topology allows optimizing the buffering and the inter-layer communication throughput, limiting the area overhead.

Our NN accelerator, called Neuro-Corgi, thus takes the form of a pipelined computational architecture, in which each layer of the network is instantiated into a specialized, parameterizable sub-architecture. These sub-architectures are then connected according to the network

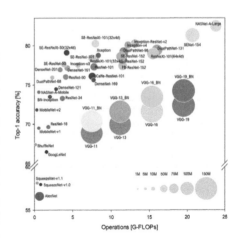

Fig. 5. Comparison of several NN topologies, as function of number of operations (X-axis), classification accuracy (Y-axis) and number of parameters (size of the circle) [9].

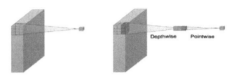

Fig. 6. (a) Standard convolution; (b) Depth-wise + point-wise convolution.

topology and parameterized to perform the inference calculation (conv, FC ...) and minimize the latency.

To simplify the architectural tradeoff analysis and the RTL generation, a back-end tool has been added to the N2D2 learning framework. This tool takes as input an algorithmic configuration file (representing the computations that need to be performed per layer) and the hardware parameters for each layer sub-architecture. It then generates files, following a 3 steps procedure: first, the generation of the topological and hardware configuration; second, the generation of the RTL code; and finally, the test and validation files.

This tool suite is very useful for architecture exploration, by varying several architectural parameters: level of

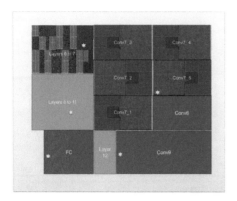

Fig. 7. NeuroCorgi initial floorplan, illustrating the placement of the different NN layers.

parallelism of each sub-architecture; size of the buffers between layers, to balance the data flow and minimize the congestions in the pipeline. ... An exploration of the design space was done by manually varying those parameters: their impact can be readily assessed at accelerator-level. The pipelined architecture allows ultra-low latency image detection (11ms). The result of initial floorplanning experiments is shown in Fig. 7.

IV. CONCLUSIONS

We aim at solving the paradox of handling ever larger image resolutions (HD) and frame rates (>30FPS), with more complex neural networks, while at the same exhibiting low latency and power values. In this work, we explored a clever, bio-inspired solution, for providing an attention mechanism to vision solutions at the edge. We focus on the dorsal stream, or "Where" path, since the "What" path is already well covered by a number of accelerators.

For pushing the energy efficiency to its maximum, several design decisions were made: a small NN topology was chosen, i.e., MobileNet-V1 to be completely integrable on-chip; weights and activations were heavily quantized (4b);

bio-inspiration was again considered, by fixing the features extraction layers (embedded memory limited to 600kB).

Our in-house learning framework N2D2 has been completed with the necessary functionalities: state-of-the-art quantization algorithms, transfer learning, hardware generation and configuration.

REFERENCES

[1] B. Dally, "CPU Computing To ExaScale and Beyond", *The International Conference for High Performance Computing, Networking, Storage, and Analysis (Super Computing)*, 2010.

[2] M. Mishkin, L.G. Ungerleider and K.A. Macko, "Object vision and spatial vision: two cortical pathways," *Trends Neuroscience*, Vol. 6, pp.414–417, 1983.

[3] V. Van Polanen and M. Davare, "Interaction between dorsal and ventral streams for controlling skilled grasp," *Neuropsychologia*, 79(Pt B), pp. 186–191, 2015.

[4] A. G. Howard *et al.*, "MobileNets: Efficient Convolutional Neural Networks for Mobile Vision Applications," https://doi.org/10.48550/arXiv.1704.04861.

[5] C. R. Siu and K. M. Murphy, "The development of human visual cortex and clinical implications," *Eye and Brain* 2018:10 25–36, doi: 10.2147/EB.S130893.

[6] https://github.com/CEA-LIST/N2D2.

[7] Q. Jin, L. Yang, A. Liao, "Towards Efficient Training for Neural Network Quantization," arXiv:1912.10207 [cs.CV], 2019.

[8] S. K. Esser *et al.*, "Learned Step Size Quantization," arXiv:1902.08153 [cs.LG], 2019.

[9] S. Bianco, R. Cadene, L. Celona and P. Napoletano, "Benchmark Analysis of Representative Deep Neural Network Architectures," in IEEE Access, vol. 6, pp. 64270-64277, 2018, doi: 10.1109/ACCESS.2018.2877890.

[10] A. Reuther *et al.*, "Survey and Benchmarking of Machine Learning Accelerators," *IEEE Conference on High Performance Extreme Computing (HPEC)*, 2019, https://doi.org/10.48550/arXiv.1908.11348.

Vincent Lorrain received his engineering degree from ESEO, Angers, France, in 2014, a M.S. degree in microelectronics from INSA, Rennes, France, in 2014, and his PhD degree in physics from Université Paris-Saclay, Orsay, France, in 2018. He has been working at CEA LIST, since 2018, where he is currently a research engineer in the development of optimized neural network architecture.

Alexandre Valentian joined CEA LETI in 2005, after an MSc and a PhD in microelectronics. His past research activities included design technology co-optimization, promoting the FDSOI technology (notably through his participation in the SOI Academy), 2.5D/3D integration technologies and non-volatile memory technology. He is currently pursuing the development of bio-inspired circuits for AI, combining memory technology, information encoding and dedicated learning methods. Since 2020, he heads the Systems-on-Chip and Advanced Technologies (LSTA) laboratory at CEA LIST.

Dr Valentian has authored or co-authored 80 conference and journal papers.

Ivan Miro-Panades received the M.S. degree in telecommunication engineering from the Technical University of Catalonia (UPC), Barcelona, Spain, in 2002, and the M.S. and Ph.D. degrees in computer science from Pierre and Marie Curie University (UPMC), Paris, France, in 2004 and 2008, respectively. He worked at Philips Research, Sureness, France and STMicroelectronics, Crolles, France, before joining CEA, Grenoble, France, in 2008, where he is currently an Expert Research Engineer in digital integrated circuits. His main research interests are artificial intelligence, the Internet of Things, low-power architectures, energy-efficient systems, and Fmax/Vmin tracking methodologies.

Inna Kucher received the M.S. degree in high-energy physics from École Polytechnique, Palaiseau, France, in 2013, and Ph.D. degree in high-energy particle physics from Université Paris-Saclay, Orsay, France, in 2017. She worked in École Polytechnique and Cern, before joining CEA LIST, in 2020, where she is currently a Research Engineer in neural networks development, optimization and deployment on embedded platforms.

Sub-mW Neuromorphic SNN Audio Processing Applications with Rockpool and Xylo

Hannah Bos and Dylan Muir

Abstract—Spiking Neural Networks (SNNs) provide an efficient computational mechanism for temporal signal processing, especially when coupled with low-power SNN inference ASICs. SNNs have been historically difficult to configure, lacking a general method for finding solutions for arbitrary tasks. In recent years, gradient-descent optimization methods have been applied to SNNs with increasing ease. SNNs and SNN inference processors, therefore, offer a good platform for commercial low-power signal processing in energy-constrained environments without cloud dependencies. Historically, these methods have not been accessible to Machine Learning (ML) engineers in industry, requiring graduate-level training to successfully configure a single SNN application. Here we demonstrate a convenient high-level pipeline to design, train and deploy arbitrary temporal signal processing applications to sub-mW SNN inference hardware. We apply a new straightforward SNN architecture designed for temporal signal processing, using a pyramid of synaptic time constants to extract signal features at a range of temporal scales. We demonstrate this architecture on an ambient audio classification task, deployed to the Xylo SNN inference processor in streaming mode. Our application achieves high accuracy (98 %) and low latency (100 ms) at low power (<100 µW dynamic inference power). Our approach makes training and deploying SNN applications available to ML engineers with general NN backgrounds, without requiring specific prior experience with spiking NNs. Our approach makes Neuromorphic hardware and SNNs an attractive choice for commercial low-power and edge signal processing applications.

Index Terms—Audio processing, Spiking Neural Networks, Deep Learning, Neuromorphic Hardware, Python.

This work was partially funded by the ECSEL Joint Undertaking (JU) under grant agreements number 876925, "ANDANTE" and number 826655, "TEMPO". The JU receives support from the European Union's Horizon 2020 research and innovation program and France, Belgium, Germany, Netherlands, Portugal, Spain, Switzerland.

Hannah Bos and Dylan Muir are with SynSense, Zürich, Switzerland (Email: dylan.muir@synsense.ai)

I. INTRODUCTION

EXISTING Deep Neural Network (DNN) approaches to temporal signal classification generally remove the time dimension from the data by buffering input windows over e.g. 40 ms and processing the entire window as a single frame [1], [2], or else apply models with complex recurrent dynamics such as Long Short-Term Memories (LSTMs) [3]. In contrast to Artificial Neural Networks (ANNs), *Spiking* Neural Networks (SNNs) include multiple temporally-evolving states with dynamics over a range of configurable time-scales. Applying these dynamics in recurrent networks forms a complex temporal basis for extracting information from temporal signals. This is achieved either through random projection [2], [4] or constructed with carefully chosen temporal properties [5]. Random recurrent architectures have historically been used for SNNs because they simplify the configuration problem — when only the readout layer is trained, configuration is performed by simply applying linear regression [4].

An alternative approach is to build feedforward networks with individual spiking units tuned to a range of various frequencies, by selecting synaptic and membrane time constants [6]. Recent advances in the optimization of SNNs using surrogate gradient descent [7], [8] have provided a feasible solution for configuring deep feedforward SNNs. However, most available libraries for simulating SNNs do not support gradient calculations, and are designed to simulate biological architectures rather than modern DNNs. At the same time, modern ML libraries for training DNNs do not support building or training SNNs.

We here demonstrate a modern ML library for SNNs, "Rockpool" [9] and its application to a new SNN inference processor "Xylo" that trains and deploys a temporal signal classification task. Recently several alternative libraries for SNN-based training with Pytorch have emerged [10], [11]. However, these libraries do not support multiple computational backends for training, and do not support deployment to neuromorphic hardware.

II. AN AMBIENT AUDIO SCENE CLASSIFICATION TASK

Audio headsets, phones, hearing aids and other portable audio devices often use noise reduction or sound shaping to improve listening performance for the user. The parameters used for noise reduction may depend on the noise level and characteristics surrounding the device and user. For example, optimal noise filtering may differ depending on whether the user is in a quiet office environment, on a street with passing traffic, or in a busy cafe with surrounding conversation.

To choose from and steer preconfigured noise reduction approaches, we propose a low-power solution to automatically and continuously classify the noise environment surrounding the user. We train and deploy an SNN on a low-power neuromorphic inference processor to perform a continuous temporal signal monitoring application, with weak low-latency requirements (environments change on the scale of minutes), but hard low-energy requirements (portable audio devices are almost uniformly battery-powered).

We use the QUT-NOISE [12] background noise corpus to train and evaluate the application. QUT-NOISE consists of multiple sequential hours of ambient audio scene recordings, from which we used the CAFE, HOME, CAR and STREET classes.

III. A TEMPORAL SIGNAL PROCESSING ARCHITECTURE FOR SNNS

We make use of slow synaptic and membrane states provided by leaky integrate-and-fire (LIF) spiking neurons to integrate information within an SNN. The dynamics of an LIF neuron are given by

$$\dot{I}_{syn} \cdot \tau_{mem} = -I_{syn} + x(t)$$
$$\dot{V}_{mem} \cdot \tau_{syn} = -V_{mem} + I_{syn} + b$$
$$V_{mem} > \theta \rightarrow \begin{cases} z(t) = z(t) + \delta(t - t_k) \\ V_{mem} = V_{mem} - \theta \end{cases}$$

Here $x(t)$ are weighted input events, I_{syn} and V_{mem} are synaptic and membrane state variables, and $z(t)$ is the train of output spikes when V_{mem} crosses the threshold θ at event times t_k. The synaptic and membrane time constants τ_{syn} and τ_{mem} provide a way to sensitise the

LIF neuron to a particular time-scale of information.

We use a range of pre-defined synaptic time constants in a deep SNN to extract and integrate temporal information over a range of scales, which can then be classified by a spiking readout layer. The proposed network architecture is shown in Figure 1.

Single-channel input audio is pre-processed through a filter bank, which extracts the power in each frequency band, spanning 50 Hz to 8000 Hz over 16 logarithmically-spaced channels. Instantaneous power is temporally quantized to 1 ms bins, with the amplitude encoded by up to 15 events per bin per channel.

Input signals are then processed by three spiking layers of 24 LIF spiking neurons in each layer, interposed with dense weight matrices. Each layer contains a fixed common τ_{mem} of 2 ms, and a range of τ_{syn} from 2 ms to 256 ms. The maximum synaptic time constant increases with each layer, such that early layers have only short τ_{syn} while final layers contain the full range of τ_{syn} values. The first layer contains neurons with two time constants of $\tau_{syn} = 2$ and 4 ms. The final layer contains neurons with $\tau_{syn} = 2, 4, 8, 16, 32, 64, 128$ and 256 ms.

The readout layer consists of four spiking neurons corresponding to the four ambient audio classes.

This network uses no bias parameters.

IV. ROCKPOOL: AN OPEN-SOURCE PYTHON LIBRARY FOR TRAINING AND DEPLOYING DEEP SNNs

Rockpool [9] is a high-level machine-learning library for spiking NNs, designed with a familiar API similar to other industry-standard python-based NN libraries. The API is similar to PyTorch [13], and in fact, PyTorch classes can be used seamlessly within Rockpool. Rockpool has the goal of making supervised training of SNNs as convenient and simple as training ANNs. The library interfaces with multiple back-ends for accelerated training and inference of SNNs, currently supporting PyTorch [13], Jax [14], Numpy, Brian 2 [15] and NEST [16], and is easily extensible. Rockpool enables hardware-aware training for neuromorphic processors, and provides a convenient interface for mapping, deployment and inference on SNN hardware from a high-level Python API.

Rockpool is installed with "pip" and "conda", and documentation is available from https://rockpool.ai. Rockpool is an open-source package, with public development based at https://github.com/syn sense/rockpool.

V. DEFINING THE NETWORK ARCHITECTURE

The network architecture shown in Figure 1 is defined in a few lines of Python code, shown in Listing 1.

VI. TRAINING APPROACH

We trained the SNN on segments of 1s duration using BPTT and surrogate gradient descent [7], [8]. We applied a mean-squared-error loss to the membrane potential of the readout neurons, with a high value for the target neuron V_{mem} and a low value for non-target neuron V_{mem}. After training, we set the threshold of the readout neurons such that the target neurons emit events for their target class and remain silent for non-target classes. Pytorch Lightning [17] was used to optimize the model against the training set using default optimization parameters.

VII. XYLO DIGITAL SNN ARCHITECTURE

We deployed the trained model to a new digital SNN inference ASIC

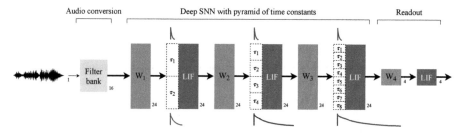

Fig. 1. Spiking network architecture for temporal signal processing. A filter bank splits single-channel audio into sixteen channels, spanning 50 Hz to 8000 Hz. The power in each frequency band is quantized to 4 bits, then injected into the SNN. The spiking network consists of three hidden layers, with a pyramid of time constants from slow to fast distributed over 24 neurons in each layer. Each layer contains several time constants, with the first hidden layer containing only short time constants (τ_1, τ_2), and the final hidden layer containing short to long time constants (τ_1 to τ_8). Finally, the readout layer outputs a continuous one-hot event-coded prediction of the current ambient audio class.

```
from rockpool.nn.combinators import Sequential
from rockpool.nn.modules import LinearTorch,
    LIFTorch
from rockpool.parameters import Constant

Nh = 24 # - Hidden layer size

# - Define pyramid of time constants over SNN
    layers
taus = [2**n * 1e-3 for n in range(1, 9)]
tau_layer1 = [taus[i] for i in range(2) for _ in
    range(Nh // 2)]
tau_layer2 = [taus[i] for i in range(4) for _ in
    range(Nh // 4)]
tau_layer3 = [taus[i] for i in range(8) for _ in
    range(Nh // 8)]

# - Define the network as a sequential list of
    modules
net = Sequential(
    LinearTorch((16, Nh)), # - Linear weights,
        hidden layer 1
    LIFTorch(Nh, tau_syn=Constant(tau_layer1)),
        # - LIF layer

    LinearTorch((Nh, Nh)), # - Hidden layer 2
    LIFTorch(Nh, tau_syn=Constant(tau_layer2)),

    LinearTorch((Nh, Nh)), # - Hidden layer 3
    LIFTorch(Nh, tau_syn=Constant(tau_layer3)),

    LinearTorch((Nh, Nh)), # - Readout layer
    LIFTorch(4) )
```

Listing 1. **Define an SNN architecture in Rockpool.** The network here corresponds to Fig 1.

"Xylo". Xylo is an all-digital spiking neural network ASIC, for efficient simulation of spiking leaky integrate-and-fire neurons with exponential input synapses. Xylo is highly configurable and supports individual synaptic and membrane time-constants, thresholds and biases for each neuron. Xylo supports arbitrary network architectures, including recurrent networks, for up to 1000 neurons. More information about Xylo can be found at https://rockpool.ai/devices/xylo-overview.html.

Figure 2 shows the logical architecture of the network within Xylo. Xylo contains 1000 LIF neurons in a hidden population and 8 LIF neurons in a readout population. Xylo provides dense input and output weights, and sparse recurrent weights with a fan-out of up to 32 targets per hidden neuron. Inputs (16 channels) and outputs (8 channels) are asynchronous firing events. The Xylo ASIC permits a range of clock frequencies, with a free choice of network time step dt.

Figure 3 shows the design of the digital LIF neurons on Xylo. Each neuron maintains independent 16-bit synaptic and membrane states. Up to 31 spike events can be generated by each neuron on each time-step if the threshold is exceeded multiple times. Each hidden layer neuron supports up to two synaptic input states. Each neuron has independently configurable synaptic and membrane time constants, thresholds, and biases. Synaptic and membrane state decay is simulated using a bit-shift approximation to an exponential decay (Listing 2). Time constants τ are converted to decay parameters *dash*, with $dash = \log_2(\tau/dt)$.

```
def bitshift(value: int, dash: int) -> int:
    new_value = value - (value >> dash)
    if new_value == value:
        new_value -= 1
    return new_value
```

Listing 2. **Python code demonstrating the bit-shift decay algorithm.** The decay parameter is given by $dash = \log_2(\tau/dt)$.

Rockpool includes a bit-accurate simulation of the Xylo architecture, "XyloSim", fully integrated with the high-level Rockpool API.

VIII. MAPPING AND DEPLOYMENT TO XYLO

a) Mapping: Rockpool provides full integration with Xylo-family hardware development kits (HDKs), supporting deployment of arbitrary network architectures to Xylo. The ability for Xylo to implement recurrent connectivity within the hidden population permits arbitrary network architectures to be deployed. Feedforward, recurrent and residual SNN architectures are all equally supported for deployment. This is accomplished by embedding feedforward

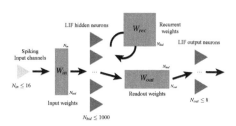

Fig. 2.   **Architecture of the digital spiking neural network inference processor "Xylo".** Xylo supports 1000 digital LIF neurons, 16 input and 8 output channels. Recurrent weights with restricted fan-out of up to 32 targets per neuron can be used to map deep feed-forward networks to the Xylo architecture.

Fig. 3.   **Digital LIF neurons on Xylo.** Each neuron maintains an integer synaptic and membrane state, with independent parameters per neuron. Exponential state decay is simulated with a bit-shift decay approach, shown in Listing 2.

network weights as sub-matrices within the recurrent weights of Xylo. Figure 4 illustrates this mapping for the network architecture of Figure 1.

b) Quantization: Floating-point parameter values must be converted to the integer representations on Xylo. For weights and thresholds, this is accomplished by considering all input weights to a neuron, then computing a scaling factor such that the maximum absolute weight is mapped to ±128, with the threshold scaled by the same factor, then rounding parameters to the nearest integer.

The deployment process is shown in Listing 3.

```
# – Extract the computational graph from a
     trained network
graph = net.as_graph()

# – Map the computational graph to the Xylo
     architecture
#   Performs DRC, assignment of HW resources,
     linearising all parameters
from rockpool.devices import xylo
spec = xylo.mapper(graph)

# – Quantize the specification using per–channel
     quantization
from rockpool.transform import
     quantize_methods as Q
spec_Q = Q.channel_quantize(**spec)

# – Deploy to a Xylo HDK
config = xylo.config_from_specification(**spec))
net_xylo = xylo.XyloSamna(hdk, config)

# – Perform inference on the HDK
output, _, _ = net_xylo(inputs)
```

Listing 3. **Mapping, quantizing and deploying a trained network to the Xylo HDK.**

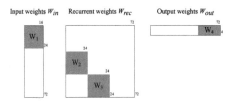

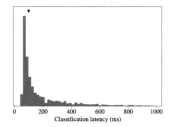

Fig. 4.   **Feedforward weights mapped to Xylo architecture.** The result of mapping the network in Figure 1 to Xylo is indicated, with dimensions and locations of sub-matrices within the Xylo architecture weights. Weight sub-matrices are labeled corresponding to the weight blocks in Figure 1.

Fig. 5.   **Distribution of correct classification latency.** Triangle: median latency of 100 ms.

IX. RESULTS

The accuracy for the trained model is given in Table I. The quantized model was deployed to a Xylo HDK, and tested on audio segments of 60 s duration. We observed a drop in accuracy of 0.8 % from the training accuracy, and a drop of 0.7 % due to model quantization.

We measured the real-time power consumption of the Xylo ASIC running at 6.25 MHz while processing test samples (Table II). Audio pre-processing ("Filter bank" in Figure 1) was performed in simulation, while SNN inference was performed on the Xylo device. We observed an average total power consumption of 542 μW while performing the audio classification task. The idle power of the SNN inference core was 219 μW, with a dynamic inference cost of 93 μW. The IO power consumption used to transfer preprocessed audio to the SNN was 230 μW. Note that in a deployed application, audio pre-processing would be performed on the device, with a concomitant reduction of IO power requirements.

Our model performs streaming classification of ambient audio with a median latency of 100 ms. Figure 5 shows the response latency distribution, from the onset of an audio sample until the first spike from the correct class output neuron.

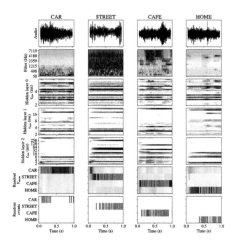

Fig. 6.   **Audio classification results on audio samples for each class (columns).** From top to bottom raw audio waveforms for each class (class indicated at the top); Filter bank outputs from low to high frequencies (indicated at left); Hidden layer responses, grouped by synaptic time constant (indicated at left); Membrane potentials V_{mem} for each of the readout neurons; and spiking events of the readout neurons (classes indicated at left).

TABLE I
AMBIENT AUDIO SCENE CLASSIFICATION
ACCURACY

Four-class accuracy (training set)	98.8 %
Validation accuracy (simulation; quantized)	98.7 %
Test accuracy (Xylo HW; 60 s samples; quantized)	98.0 %

Figure 6 shows several examples of audio samples classified by the trained network.

a) Inference energy benchmarks: Our network performs continuous non-frame-based inference, making a precise

TABLE II
CONTINUOUS POWER MEASUREMENTS

SNN core idle power (Xylo HDK)	219 μW
SNN core dynamic inference power (Xylo HDK; 60 s samples)	93 μW
SNN core total inference power (Xylo HDK; 60 s samples)	**312 μW**
Total IO power (Xylo HDK; 60 s samples)	230 μW
Total inference power (Xylo HDK; 60 s samples)	**542 μW**

TABLE III
PER-INFERENCE ENERGY MEASUREMENTS

Inference rate (med. latency)	10 Hz
Total energy per inference (med. latency)	**54.2 μJ**
Total energy per network time-step	542 nJ
Dynamic energy per inference (med. latency)	**9.3 μJ**
Dynamic energy per network time-step	93 nJ

TABLE IV
PER-NEURON PER-INFERENCE ENERGY
COMPARISON

Citation	Device	N	E_{Tot}/N	E_{Dyn}/N
[2]	Quadro K4000	512	95.9 μJ	58.0 μJ
[2]	Xeon E5-2630	512	30.7 μJ	12.4 μJ
[2]	Jetson TX1	512	23.2 μJ	10.9 μJ
[18]	Cortex M4F	2176	—	5.15 μJ
[2]	Movidius NCS	512	4.21 μJ	2.85 μJ
[2]	Loihi	512	0.73 μJ	0.53 μJ
[18]	MAX78000	2176	—	**0.12 μJ**
This work	Xylo-A2[†]	76	**0.71 μJ**	**0.12 μJ**

N: Number of neurons. E_{Tot}: Total energy per inference.
E_{Dyn}: Dynamic energy per inference. [†]Based on med. latency of 100 ms.

definition of "an inference" complicated. We considered two possible definitions for inference time: one based on the median latency (100 ms; Figure 5); and one based on the time taken to perform a full evaluation of the network (network time-step of 1 ms). Based on the continuous power measurements in Table II, our system exhibits per-inference dynamic energy consumption of 9.3 μJ (med. latency) and 93 nJ (network time-step). Per-inference total energy consumption was 54.2 μJ (med. latency) and 542 nJ (network time-step). These results are summarised in Table IV.

Recent work deploying a keyword-spotting application to low-power CNN inference hardware achieved total energy consumption of 251 μJ per inference on the optimized Maxim MAX78000 accelerator, and 11 200 μJ per inference on a low-power microprocessor (ARM Cortex M4F) [18]. This corresponded to a continuous power consumption of 71.7 mW (MAX78000) and 12.4 mW (Cortex M4F) respectively.

Previous work benchmarking audio processing applications with regard to power consumption compared desktop-scale inference with spiking neuromorphic hardware [2]. In a keyword spotting task, dynamic energy costs ranged from 0.27 mJ to 29.8 mJ per inference, covering spike-based neuromorphic hardware (intel Loihi) to a GPU device (Quadro K4000). This corresponded to a range of continuous power consumption from 0.081 W to 22.86 W.

Published numbers for the mixed-signal low-power audio processor

Syntiant NDP120 place the device at 35 μJ to 50 μJ per inference on a keyword spotting task, with continuous power consumption of 8.2 mW to 28 mW [19].

Dynamic inference energy scales roughly linearly with the number of neurons in a network [2]. In Table IV we report a comparison between auditory processing tasks on various architectures, normalized by network size. At 0.12 μJ dynamic inference power per inference per neuron, our implementation on Xylo requires equal low energy with the MAX78000 CNN inference processor. However the MAX78000 CNN core requires 792 μW to 2370 μW in inactive mode [20], compared with 219 μW for Xylo, making Xylo more energy efficient in real terms.

X. CONCLUSION

We demonstrated a general approach for implementing audio processing applications using spiking neural networks, deployed to a low-power Neuromorphic SNN inference processor "Xylo". Our solution reaches high accuracy (98 %)

with <100 spiking neurons, operating in streaming mode with low latency (med. 100 ms) and at low power (<100 μW dynamic inference power). Xylo exhbits lower idle power, lower dynamic inference power and lower energy per inference than other low-power audio processing implementations.

Our software pipeline "Rockpool" (rockpool.ai) provides a modern Machine Learning approach to building applications, with a convenient high-level API for defining neural network architectures. Rockpool supports the definition and training of SNNs via several automatic differentiation back-ends. Rockpool also supports quantization, mapping, and deployment to SNN inference hardware with a few lines of Python code.

Our approach supports commercial design and deployment of SNN applications, by making the configuration process of SNNs accessible to ML engineers without graduate-level training in SNNs.

Here we have not demonstrated the full capabilities of Rockpool, which also supports residual spiking architectures, quantization- and hardware-aware training, training for time constants and other neuron parameters, and high extensibility for additional computational back-ends.

We anticipate SNNs and low-power neuromorphic inference processors to contribute significantly to the current push for low-power machine learning at the edge.

REFERENCES

[1] G. Chen, C. Parada, and G. Heigold, "Small-footprint keyword spotting using deep neural networks," in *2014 IEEE International Conference on Acoustics, Speech and Signal Processing (ICASSP)*, 2014, pp. 4087–4091.

[2] P. Blouw, X. Choo, E. Hunsberger, and C. Eliasmith, "Benchmarking keyword spotting efficiency on neuromorphic hardware," in *Proceedings of the 7th Annual Neuro-Inspired Computational Elements Workshop*, ser. NICE '19. New York, NY, USA: Association for Computing Machinery, 2019. [Online]. Available: https://doi.org/10.1145/3320288.3320304

[3] J. Deng, B. Schuller, F. Eyben, D. Schuller, Z. Zhang, H. Francois, and E. Oh, "Exploiting time-frequency patterns with LSTM-RNNs for low-bitrate audio restoration," *Neural Computing and Applications*, vol. 32, no. 4, pp. 1095–1107, Feb. 2020. [Online]. Available: https://doi.org/10.1007/s00521-019-04158-0

[4] W. Maass and H. Markram, "On the computational power of circuits of spiking neurons," *Journal of Computer and System Sciences*, vol. 69, no. 4, pp. 593–616, 2004. [Online]. Available: https://www.sciencedirect.com/science/article/pii/S0022000004000406

[5] A. Voelker, I. Kajić, and C. Eliasmith, "Legendre memory units: Continuous-time representation in recurrent neural networks," in *Advances in Neural Information Processing Systems*, H. Wallach, H. Larochelle, A. Beygelzimer, F. d'Alché-Buc, E. Fox, and R. Garnett, Eds., vol. 32. Curran Associates, Inc., 2019. [Online]. Available: https://proceedings.neurips.cc/paper/2019/file/952285b9b7e7a1be5aa7849f32ffff05-Paper.pdf

[6] P. Weidel and S. Sheik, "Wavesense: Efficient temporal convolutions with spiking neural networks for keyword spotting," 2021. [Online]. Available: https://arxiv.org/abs/2111.01456

[7] J. H. Lee, T. Delbruck, and

M. Pfeiffer, "Training deep spiking neural networks using backpropagation," *Frontiers in Neuroscience*, vol. 10, 2016. [Online]. Available: https://www.frontiersin.org/articles/10.3389/fnins.2016.00508

[8] E. O. Neftci, H. Mostafa, and F. Zenke, "Surrogate gradient learning in spiking neural networks: Bringing the power of gradient-based optimization to spiking neural networks," *IEEE Signal Processing Magazine*, vol. 36, no. 6, pp. 51–63, 2019.

[9] D. Muir, F. Bauer, and P. Weidel, "Rockpool documentaton," Mar 2019.

[10] C.-G. Pehle and J. E. Pedersen, "Norse — A deep learning library for spiking neural networks," Jan. 2021, documentation: https://norse.ai/docs/. [Online]. Available: https://doi.org/10.5281/zenodo.4422025

[11] J. K. Eshraghian, M. Ward, E. Neftci, X. Wang, G. Lenz, G. Dwivedi, M. Bennamoun, D. S. Jeong, and W. D. Lu, "Training spiking neural networks using lessons from deep learning," 2021. [Online]. Available: https://arxiv.org/abs/2109.12894

[12] D. Dean, S. Sridharan, R. Vogt, and M. Mason, "The QUT-NOISE-TIMIT corpus for evaluation of voice activity detection algorithms," in *Proceedings of the 11th Annual Conference of the International Speech Communication Association.* International Speech Communication Association, 2010, pp. 3110–3113.

[13] A. Paszke, S. Gross, F. Massa, A. Lerer, J. Bradbury, G. Chanan, T. Killeen, Z. Lin, N. Gimelshein, L. Antiga, A. Desmaison, A. Kopf, E. Yang, Z. DeVito, M. Raison, A. Tejani, S. Chilamkurthy, B. Steiner, L. Fang, J. Bai, and S. Chintala, "PyTorch: An imperative style, high-performance deep learning library," in *Advances in Neural Information Processing Systems 32.* Curran Associates, Inc., 2019, pp. 8024–8035. [Online]. Available: http://papers.neurips.cc/paper/9015-pytorch-an-imperative-style-high-performance-deep-learning-library.pdf

[14] J. Bradbury, R. Frostig, P. Hawkins, M. J. Johnson, C. Leary, D. Maclaurin, G. Necula, A. Paszke, J. VanderPlas, S. Wanderman-Milne, and Q. Zhang, "JAX: composable transformations of Python+NumPy programs," 2018. [Online]. Available: http://github.com/google/jax

[15] M. Stimberg, R. Brette, and D. F. Goodman, "Brian 2, an intuitive and efficient neural simulator," *eLife*, vol. 8, p. e47314, Aug. 2019.

[16] M.-O. Gewaltig and M. Diesmann, "NEST (NEural Simulation Tool)," *Scholarpedia*, vol. 2, no. 4, p. 1430, 2007.

[17] W. Falcon et al., "PyTorch Lightning," *GitHub*, 2019. [Online]. Available: https://github.com/PyTorchLightning/pytorch-lightning

[18] M. G. Ulkar and O. E. Okman, "Ultra-low power keyword spotting at the edge," 2021. [Online]. Available: https://arxiv.org/abs/2111.04988

[19] "MLPerf™ v0.7 Inference: Tiny Keyword Spotting, entries 0.7-2012 and 0.7-2013. result verified by MLCommons Association. The MLPerf™ name and logo are trademarks of MLCommons Association in the United States and other countries. All rights reserved. Unau-

thorized use strictly prohibited. See www.mlcommons.org for more information." 2021. [Online]. Available: https://mlcommons.org/en/inference-tiny-07/

[20] "MAX78000 artificial intelligence microcontroller with ultra-low-power convolutional neural network accelerator," May 2021. [Online]. Available: https://www.maximinteg rated.com/en/products/microcontrol lers/MAX78000.html

Hannah Bos Dr. Bos is a Senior Algorithms and Applications ML Engineer at SynSense, with a background in computational Neuroscience and theoretical Physics. At SynSense she designs algorithms for neuromorphic chips and supports the development of new hardware. Dr. Bos holds a Ph.D. in Physics and theoretical Neuroscience from RWTH Aachen, and a Masters in Physics from the University of Oslo. She researched computational neuroscience at the University of Pittsburgh.

Dylan Muir Dr. Muir is the Vice President for Global Research Operations; Director for Algorithms and Applications; and Director for Global Business Development at SynSense. Dr. Muir is a specialist in architectures for neural computation. He has published extensively in computational and experimental neuroscience. At SynSense he is responsible for the company research vision, and directing the development of neural architectures for signal processing. Dr. Muir holds a Doctor of Science Ph.D. from ETH Zürich, and undergraduate degrees (Masters) in Electronic Engineering and Computer Science from QUT, Australia.

An Embedding Workflow for Tiny Neural Networks on Arm Cortex-M0(+) Cores

Jianyu Zhao, Cecilia Carbonelli, and Wolfgang Furtner, *Infineon Technologies AG*

Abstract—Neural networks are becoming increasingly widely used in always-on IoT edge devices for more precise and secure data analysis with less latency. However, due to the strict cost and power constraints for such applications, only the smallest microcontrollers, typically equipped with Arm Cortex-M0(+) cores, could be used for algorithm implementation. For a memory of only a few tens of kilobytes, the available open-source embedding tools either are too large or require too much handcraft effort. In this paper, we propose an end-to-end embedding workflow focused on tiny neural network deployment on Arm Cortex-M0(+) cores. The method covers all the steps, including network quantisation, C code generation and performance verification. A Python and C library was developed following the proposed method and validated on a low-cost environmental sensing application. As a result, up to 73.9% of the memory footprint could be reduced with the quantised network with only a small sacrifice in performance. While reducing the manual effort of network embedding to the minimum, the workflow remains flexible enough to allow for customisable bit shifts and different layer combinations.

Index Terms—Arm Cortex-M0(+), environmental sensing, network quantisation

Jianyu Zhao, Cecilia Carbonelli, and Wolfgang Furtner are with Infineon Technologies AG, 85579 Neubiberg, Germany. The corresponding author is J. Zhao (jianyu.zhao@infineon.com). The research leading to these results has received funding from the European Union's ECSEL Joint Undertaking under grant agreement n° 826655 - project TEMPO.

I. INTRODUCTION

IN the past decade, the availability of large amounts of data and the ever-increasing computing power have enabled the explosive development of machine learning algorithms, especially neural networks. Meanwhile, growing concerns about data privacy have drawn people's attention from centralised cloud computing to more distributed edge implementation. Consequently, there is an increasing interest in the use of neural networks for more secure and precise on-sensor data analysis and the joint optimisation of the algorithm and the dedicated edge hardware, such as microcontrollers. Due to the cost and power constraint for such applications, the hardware platform is often so small, with a memory of only a few tens of kilobytes, that the implementation has to run on bare metal, without the support of an operating system. In this work, we propose an end-to-end embedding workflow for the quantisation of small neural networks, the generation of C source files and the accuracy evaluation. Focusing on Arm Cortex-M0(+), the workflow provides unparalleled simplicity while keeping fine-tuning possibilities.

II. BACKGROUND

Artificial neural networks have existed for decades [1], but it was not until

2012, when AlexNet won the ImageNet Large Scale Visual Recognition Challenge by using graphic processing units (GPUs) during training [2], that the algorithm gained momentum and attracted increased attention from academia and industry. Inspired by the biological neural networks in animal brains, artificial neural networks consist of multiple layers of inter-connected computing units, called "neurons", which are essentially matrix multiplications and non-linear activation functions. Such networks can model complicated non-linear mappings without humans manually choosing the parameters.

Besides image processing, with the growth of the Internet of Things (IoT) technologies, it is increasingly popular to use neural networks also for the analysis of data from other types of sensors, such as microphones [3, 4] radars [5] and even gas sensors [6]. Compared with most handcrafted algorithmic models, neural networks are capable of modelling more complex relations and thus delivering much closer fitting for many use cases. Although the training process with back-propagation is computationally expensive, the inference process can boil down to loops of simple multiplications and accumulations, making it possible to run such models on restricted platforms.

Different from conventional neural network applications, such as image and natural language processing, sensor applications are often much more constrained w.r.t. power consumption, material cost and footprint. Such devices are often always-on and most likely battery-powered, so the power limit is typically under 1 mW [7], while a mobile or desktop CPU needs Watts. Cost also plays an important role in the selection of processors. As the smallest member of the Arm Cortex-M series, Cortex-M0(+) costs less than a euro each, enables developers to

achieve 32-bit performance at an 8-bit price point [8, 9] and thus has become the state-of-the-art for IoT edge devices. As many applications already have an M0 core for simple sensor control, implementing a machine learning algorithm on the same core would create additional value without increasing the material cost. To further reduce the footprint and power consumption of the whole system, a system in a package (SiP) [10] could be used to stack the microcontroller and the sensors in one single package.

However, with opportunities, there also come challenges. Small platforms such as Arm Cortex-M0(+) do not support any operating system or file system, so the code needs to run on bare metal and the network parameters need to be saved as C source files. The memory resource is also minimal, typically 16 or 32 kB Flash and 4 or 8 kB SRAM, making most available machine learning frameworks too big for the platform. In addition, the lack of a floating-point unit (FPU) also makes floating-point arithmetic extremely expensive. However, most neural networks are trained with single-precision floating-points inputs and parameters.

Fortunately, more deep learning frameworks are being developed for different hardware platforms. Arm, for example, has extended its Common Microcontroller Software Interface Standard (CMSIS) with a dedicated neural network library called CMSIS-NN [11], targeting Cortex-M processors with specific networks. However, the overall process from quantisation to embedding still remains manual, and the developers need to spend a lot of time coming up with the quantisation solution, writing C source code with the library, and then evaluating the performance of the chosen quantisation settings. TensorFlow Lite for Microcontroller (TF Lite Micro) [12] is another promising solution for the Cortex-M series, as it is seamlessly integrated into the popular TensorFlow framework for model training, and one can convert the original model to Flatbuffer format and then generate embedded source code with just a few lines of Python code. As an extension of TensorFlow Lite that targets larger platforms such as the Cortex-A series, TF Lite Micro libraries are also written in C++ 11. However, the most popular language for embedded programming is C, and C++ 11 is still not supported by toolchains available on many platforms. Also, the generated models are serialised as arrays of bytes, which is hard to interpret and customise implementation details. Note that although the network embedding could be some extent automated, the other parts of the embedded firmware, *e.g.,* pre-processing and feature extraction, still need to be implemented manually in C. Therefore, it is crucial that the network implementation is as transparent as possible.

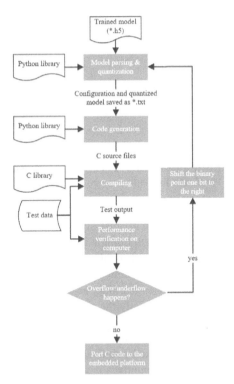

Fig. 1. An overview of the embedding workflow proposed in this work.

III. METHODS

The training of neural network models is often carried out on computers or computer clusters with GPUs. With various machine learning frameworks available on these platforms, algorithm developers can focus on the architecture design without worrying about the underlying implementation. The trained models are often Python objects with properties describing the network architecture and the trained parameters as 32-bit floating-point values.

For the implementation of a model on an Arm Cortex-M0(+) platform, a workflow is suggested in Fig. 1 with a library pre-written in Python and C.

A. Preparation

When it comes to network quantisation, there are generally two approaches:

quantisation-aware training [13] and post-training quantisation [14]. In this work, we use the latter because it is more flexible, doesn't require retraining and could achieve reasonably good performance in most cases [14].

Besides trained models, some test data with the corresponding target values are also needed for performance verification.

B. Quantisation

As the platform does not have an FPU, floating-point arithmetic can be very expensive. By converting the model parameters from 32-bit floating-point values to 16- or 8-bit integers, it is possible to reduce the memory footprint and computational cost simultaneously without sacrificing much accuracy.

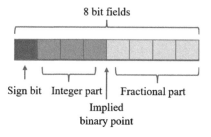

Fig. 2. A signed 8-bit fixed-point representation of a fractional number, with the binary point positioned after the 4th bit.

Quantisation is generally defined as the process of constraining a large set of values (such as real numbers) to a discrete set (such as the integers) [15]. In the context of deep learning, network quantisation is essentially the process of representing model parameters with fewer bits, *e.g.*, 8 bits, as mentioned above.

An 8-bit fixed-point representation of the fractional numbers is shown in Fig. 2. In this example, 4 bits are used for the integer part, while the other 4 bits are for the fractional part. The first bit of the integer part is used as a sign bit so that it is possible to represent both positive and negative values.

The choice of the binary point position is crucial when we quantise decimals. In most cases, two factors need to be considered:

- the dynamic range of the variable, depending on the lowest and highest value that one needs to represent within the given algorithm;
- the highest tolerable quantisation error.

Ideally, after calculating the least number of bits needed to represent the full range of the model parameters, the developer could use the remaining bits for the fractional part. However, during the multiplication and accumulation operations of the neural networks, overflow or underflow can still happen when the dynamic range exceeds the static one calculated from the parameters. With the workflow described in Fig. 1, we suggest first quantising the parameters with as many fractional bits as possible and verifying if overflow or underflow happens with the output of the C implementation before exporting the source code to the target platform. When it does happen, we can go back to quantisation and shift the binary point one bit to the right until the whole dynamic range with the test data is covered.

After quantisation, the network architecture is saved in a configuration text file, and the quantised parameters are saved in another file, one layer after another. These files would be parsed in

```
4
Dense 4 8 hard_sigmoid
Dense 8 3 softmax
4 3
```

Fig. 3. Example of the configuration txt file for a classifier trained for the iris dataset. The model has four inputs, three outputs and two dense layers, each with a different activation function.

```
6
5
16 6 -1 7 -19 -1 6 13
-46 -49 41 -45 -24 6 73 53
68 79 -78 69 -65 -87 -84 -91
74 95 -80 68 -99 -124 -71 -105
8
-26
4
23
44
67
-32
43
-84 7 50
-78 -17 37
75 13 -50
-82 7 35
8 43 -63
38 47 -46
73 -10 -42
66 31 -63
3
3
-9
```

Fig. 4. Example of the txt file for the quantised 8-bit parameters from the model mentioned in Fig. 3. The 6 and 5 at the beginning represent the numbers of bits used for the fractional part of the numbers for the first and the second dense layers, respectively. They are followed by weights and biases quantised accordingly.

the next step to generate C source files. Examples of the configuration file and the parameter file are given in Fig. 3 and Fig. 4.

C. library and code generation

Before generating any C source code for a specific model, a general-purpose library needs to be prepared to perform the basic layer tasks, as shown in Fig. 5. It is recommended to start with dense and Gated Recurrent Unit (GRU) layers, as they are often used in time series analyses for sensor data.

Pre-defining the data structures for different types of layers and different variable lengths, the library also provides standardised interfaces for all the layers supported. Following the mathematical definition of the specific layers, the detailed implementation boils down to a few loops with multiply and accumulate operations. An example of the 8-bit dense layer interface is provided in Fig. 6. Other layers, 16-bit dense, 8-bit GRU and 16-bit GRU, have similar designs.

As no MAC (multiply and accumulation) is available on Cortex-M0(+) cores, multiplications and accumulations are calculated with basic instructions with bit shifts.

With all the implementations available for each layer of the model, it is already possible for the developer to handcraft a C solution. However, copying all the parameters from txt files and reformatting

```
typedef struct {
    const int8_t **weights;
    const int8_t *biases;
    const uint8_t activation;
    const uint8_t dim_out;
    const uint8_t shift;
} dense_param_8bit_t;
```
(a)

```
void dense_8bit(
    int16_t *input,
    const dense_param_8bit_t *params,
    uint8_t dim_in,
    const uint8_t input_shift,
    int16_t *output);
```
(b)

Fig. 6. (a) The C structure to save 8-bit parameters for a dense layer, including a double pointer for the 2d weight array, a pointer to the 1d bias array, an unsigned 8-bit integer to suggest the type of activation function, an unsigned 8-bit integer for the size of the layer output and another for the bit shift. (b) The corresponding implementation of an 8-bit dense layer, which besides layer parameters also takes a pointer to a 1d array for the input values and modifies the array pointed by the output pointer in return.

them in complicated C structures with pointers and double pointers can be time-consuming and error-prone. In this paper, we propose using a standardised Python script, which reads the model parameters in the txt file following the network architecture given in the configuration file and saves them in the desired format in a C header file. The implementation is written in the corresponding C file. An example of the generated code for an 8-bit 3-hidden-layer classifier is given in Fig. 7.

Note that the model parameters are declared as constant variables to save SRAM usage.

Meanwhile, the main function is also generated to take the test input data from a txt file, feed the samples one by one to the C model function and export the output in a txt file for verification.

D. Performance evaluation in Python

The exported C output can be loaded and compared against the output provided by the original Python model. If the results are drastically different, for

```
Name
----
activation_functions.c
activation_functions.h
fixed_point_operations.c
fixed_point_operations.h
nn_helpers.c
nn_helpers.h
nn_layers.c
nn_layers.h
softmax_functions.c
softmax_functions.h
```

Fig. 5. Header and C files from the C library.

```
typedef struct {
    const dense_param_8bit_t *layer1;
    const dense_param_8bit_t *layer2;
    const uint8_t n_features;
    const uint8_t n_out;
} model_param_t;
```

(a)

```
static const model_param_t model_params = {
    &(const dense_param_8bit_t)
    {
        (const int8_t *[])
        {
            (const int8_t []) {16, 6, -1, 7, -19,
-1, 6, 13},
            (const int8_t []) {-46, -49, 41, -45,
-24, 6, 73, 53},
            (const int8_t []) {68, 79, -78, 69,
-65, -87, -84, -91},
            (const int8_t []) {74, 95, -80, 68,
-99, -124, -71, -105}
        },
        (const int8_t []) {8, -26, 4, 23, 44, 67,
-32, 43},
        ACTIVATION_SIGMOID,
        // dim_out, shift
        8, 6
    },
    ...
    4, 3
}
```

(b)

```
int16_t z1[model_params->layer1->dim_out];

dense_8bit(features, model_params->layer1,
    model_params->n_features, input_shift, z1);
dense_8bit(z1, model_params->layer2,
    model_params->layer1->dim_out,
    model_params->layer1->shift, pred);
```

(c)

Fig. 7. Example of the generated C source code. (a) Definition of a comprehensive parameter structure, which consists of a pointer for the parameter structure of each of the hidden layers and two integers for the dimensions of the network input and output. (b) Declaration of the model parameter (only the first layer is shown due to space limitations) with automatically filled model parameters read from the parameter text file illustrated in Fig. 3. (c) Simple network implementation with available layer implementations.

example, the C output saturates at a certain level, there could be an overflow or underflow. The developer may go back and shift the binary point one bit to the right, generate the code again and evaluate the model performance until no overflow or underflow happens again.

IV. RESULTS AND DISCUSSION

The proposed workflow has been implemented as a Python and C library and tested extensively with classification and regression networks trained with various public data sets, such as Boston housing price, Iris and MNIST. At Infineon Technologies AG, we also applied the workflow to our in-house low-cost environmental sensors and managed to enable smart gas sensing on a PSoC® analogue coprocessor, which comes with a Cortex-M0+ core. In this section, we will introduce the application and then discuss the results of the embedded algorithm.

With the growing concerns about air quality, there is also a higher demand for fine-granular sensing for many health applications [16]. Conventionally, air quality is monitored at the city level, with large monitoring stations logging the concentrations of target gases and particles every hour. Typically using laser-based spectroscopy analysis [17], such as cavity ring-down spectroscopy, the stations are accurate but take a lot of space and are expensive to set up and maintain [18, 19]. The recorded 1- or even 8-hour average concentrations of the gas pollutants also do not necessarily reflect the air quality in a local environment. With more accurate, low-cost electrochemical sensors enabled by new sensing materials and embedded neural networks, it is possible to measure the concentration of the target gas(es) with finer granularity, e.g., in a room or outdoor on a battery-powered device, offering opportunities to develop various new applications. The hardware components of such a gas-measuring platform are illustrated in Fig. 8.

For the specific gas-sensing application, which we address in this section, a small GRU network is trained to estimate gas concentrations with historical data saved in a buffer. The architecture of the example network is provided in Fig. 9. It can exploit the time properties of the sensor signals while keeping the memory footprint within budget. The algorithmic model was first designed and trained on

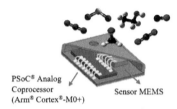

PSoC® Analog
Coprocessor
(Arm® Cortex®-M0+)   Sensor MEMS

Fig. 8. A low-cost environmental sensing platform with a neural network embedded on an Infineon PSoC® analogue Coprocessor in the same package.

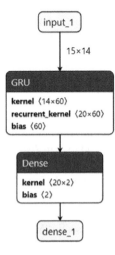

Fig. 9. Network architecture. The example network has 14 extracted features, 15 timesteps, 20 hidden units, and 2 output values.

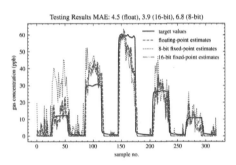

Fig. 10. Comparison of the estimated gas concentrations from the floating-point, 16-bit fixed-point and 8-bit fixed-point algorithm implementations.

a computer cluster and then deployed on an Infineon PSoC® analogue coprocessor, which is not only used for real-time concentration estimation but also signal measurement, heater control and communication.

The analogue coprocessor platform is equipped with an Arm Cortex M0+ processor with 32 kB Flash and 4 kB SRAM. Following the workflow proposed in section III, it is possible to embed the model, visualise the simulated output, flexibly adjust the position of the binary point, and thus find the best trade-off between algorithm performance and memory footprints.

The test results with a given data set are shown in Fig. 10. In the test experiment, the target gas was increased to specific concentrations and then decreased to 0 ppb. The concentration estimates of the original floating-point model are very close to the ground truth, with slight over- or underestimation for some of the samples. The results from the 16-bit fixed-point model are similar to the ones from the original, with a mean absolute error (MAE) of 3.9 ppb. A more significant quantisation error is seen with the 8-bit model. It tends to overestimate a lot, especially during the first pulse at 10 ppb, resulting in an MAE of 6.8 ppb.

The accuracies and memory footprints of the neural networks of different data types are summarised in Table I. By quantising the network to 16-bit, we saved 49.3% of the memory with little change in performance; when the

TABLE I
PERFORMANCE COMPARISON OF DIFFERENT
NETWORK IMPLEMENTATIONS

Network Implementation	Performance Metrics		
	MAE^a (ppb)	Flash (kB)	SRAM (kB)
Floating-point	4.5	33.5	1.8
16-bit fixed-point	3.9	16.7	1.2
8-bit fixed-point	6.8	8.4	0.8

[a]Mean absolute error

network is further reduced to 8-bit, another 24.6% of memory is reduced at the price of the noisy and less accurate regression results. Given the sizes of the Flash and SRAM, quantising the network to 16-bit or even 8-bit makes it possible to embed a GRU that would otherwise not fit on the platform.

V. Conclusion

In this paper, we propose an embedding workflow optimised for tiny neural networks and Arm Cortex-M0(+) cores. Without the overhead of developing a single universal solution for all the embedded platforms, it is possible to have an end-to-end solution that covers every step, from parameter quantisation, code generation to performance evaluation with one library written in Python and C. While reducing the manual effort of networking embedding to the minimum, the workflow is flexible enough for different layer combinations and customisable bit shifts. As Cortex-M0(+) is already part of many low-power sensor systems but only used for basic sensor control, values can be easily added to such products without additional material costs.

References

[1] R. Lippmann, "An introduction to computing with neural nets," in IEEE ASSP Magazine, vol. 4, no. 2, pp. 4-22, Apr 1987, doi: 10.1109/MASSP.1987.1165576.

[2] A. Krizhevsky, I. Sutskever and G. Hinton, "ImageNet classification with deep convolutional neural networks," Communications of the ACM, vol. 60, no. 6, pp. 84-90, 2017, doi: 10.1145/3065386.

[3] K. Kumatani, J. McDonough and B. Raj, "Microphone Array Processing for Distant Speech Recognition:

From Close-Talking Microphones to Far-Field Sensors," in IEEE Signal Processing Magazine, vol. 29, no. 6, pp. 127-140, Nov. 2012, doi: 10.1109/MSP.2012.2205285.

[4] M. Wu et al., "Monophone-Based Background Modeling for Two-Stage On-Device Wake Word Detection," 2018 IEEE International Conference on Acoustics, Speech and Signal Processing (ICASSP), 2018, pp. 5494-5498, doi: 10.1109/ICASSP.2018.8462227.

[5] M. Scherer, M. Magno, J. Erb, P. Mayer, M. Eggimann and L. Benini, "TinyRadarNN: Combining Spatial and Temporal Convolutional Neural Networks for Embedded Gesture Recognition With Short Range Radars," in IEEE Internet of Things Journal, vol. 8, no. 13, pp. 10336-10346, 1 July1, 2021, doi: 10.1109/JIOT.2021.3067382.

[6] X. Zhai, A. A. S. Ali, A. Amira and F. Bensaali, "MLP Neural Network Based Gas Classification System on Zynq SoC," in IEEE Access, vol. 4, pp. 8138-8146, 2016, doi: 10.1109/ACCESS.2016.2619181.

[7] P. Warden and D. Situnayake, "Chapter 1. Introduction," in TinyML, 1st ed., Sebastopol, CA, USA: O'Reilly Media, Inc, 2019, pp. 1-3.

[8] arm. (2022, Aug. 12). arm CPU Cortex-M0: Tiny, 32-bit Processor. [online]. Available: https://www.arm.com/products/silicon-ip-cpu/cortex-m/cortex-m0.

[9] arm. (2022, Aug. 12). arm CPU Cortex-M0+: 32-bit, Low-Power Processor at an 8-bit Cost. [online]. Available: https://www.arm.com/products/silicon-ip-cpu/cortex-m/cortex-m0-plus.

[10] A. Fontanelli, "System-in-Package Technology: Opportunities and

Challenges," *9th International Symposium on Quality Electronic Design (isqed 2008)*, 2008, pp. 589-593, doi: 10.1109/ISQED.2008.4479803.

[11] L. Lai, N. Suda, V. Chandra, "CMSIS-NN: Efficient Neural Network Kernels for Arm Cortex-M CPUs," *arXiv preprint,* 2018. Available: arXiv:1801.06601

[12] TensorFlow, (2022, Aug. 12). *TensorFlow Lite for Microcontrollers.* [online]. Available: https://www.tensorflow.org/lite/microcontrollers

[13] B. Jacob et al., "Quantization and Training of Neural Networks for Efficient Integer-Arithmetic-Only Inference," *2018 IEEE/CVF Conference on Computer Vision and Pattern Recognition*, 2018, pp. 2704-2713, doi: 10.1109/CVPR.2018.00286.

[14] R. Krishnamoorthi, "Quantizing deep convolutional networks for efficient inference: A whitepaper," *arXiv preprint,* 2018. Available: arXiv:1806.08342

[15] R. M. Gray and D. L. Neuhoff, "Quantization," in *IEEE Transactions on Information Theory*, vol. 44, no. 6, pp. 2325-2383, Oct. 1998, doi: 10.1109/18.720541.

[16] P. Kumar et al., "The rise of low-cost sensing for managing air pollution in cities," *Environment international 75*, 2015, pp. 199-205.

[17] W. X. Peng, K. W. D. Ledingham, A. Marshall and R. P. Singhal, "Urban air pollution monitoring: Laser-based procedure for the detection of NOx gases," *Analyst 120.10,* 1995, pp. 2537-2542.

[18] CEN Ambient Air, "Standard Method for the Measurement of the Concentration of Nitrogen Dioxide and Nitrogen Monoxide by Chemiluminescence (EN 14211:2012)," *European Committee for Standardization*, 2012.

[19] CEN Ambient Air, "Standard Method for the Measurement of the Concentration of Ozone by Ultraviolet Photometry (EN 14625:2012)," *European Committee for Standardization*, 2012.

Jianyu Zhao was born in Sichuan, China, in 1993. She received a B. Eng. (2015) from Tianjin University, China, and her M.Sc. (2018) from the Technical University of Munich, Germany, both in electrical and information engineering.

In 2017, she joined Infineon Technologies AG, Neubiberg, Germany, to write her master thesis on analysing gas sensor data using machine learning approaches. In 2018, she joined Infineon as an algorithm and modelling engineer and continued to work on algorithm development and implementation for smart sensing technologies. She holds several international patents, and her current interest is deploying neural networks on edge devices.

Cecilia Carbonelli is a Senior Principal - System and Algorithm Architect at Infineon Technologies AG. She studied telecommunications engineering and earned a PhD in information engineering from the University of Pisa in 2005. She was a post-doc at University of Southern California for a couple of years and then moved into industry and to Germany, joining Infineon in December 2006. She has been a system engineer over the last 15 years, working on cellular standards and modem platforms, physical layer algorithms, machine learning and AI applied to sensor products. She is the author of 35 scientific publications and numerous patents.

 **Wolfgang Furtner** is a Distinguished Engineer for SoC Architectures at Infineon Technologies AG. He received his degree in electrical engineering from the University of Applied Sciences Munich, Germany. He started his career working 4 years in a startup developing Graphics Processors (GPUs), followed by 11 years architecting graphics and video processing ICs at Philips Semiconductors. Since 2006 he has been with Infineon and heading System Concept Engineering for power and sensors. His interests are embedded architectures for artificial intelligence and machine learning, smart sensors and system architectures for quantum computing.

Edge AI Platforms for Predictive Maintenance in Industrial Applications

Ovidiu Vermesan and Marcello Coppola

Abstract—The use of intelligent edge-sensing devices for measuring various parameters (vibration, temperature, etc.) for industrial equipment/motors using artificial intelligence (AI), machine learning (ML), and neural networks (NNs) is being increasingly adopted in industrial predictive maintenance (PdM) applications. Developing and deploying ML algorithms and NNs on edge devices using sensors and microcontroller processing units based on Arm® Cortex®-M cores (e.g., M0, M0+, M3, M4, M7) microcontrollers requires robust AI-based platforms and workflows. This paper highlights the importance of adequately architecting AI workflow for PdM in industrial applications at the edge. New platforms have recently emerged with various degrees of automation and customisation for end-to-end development and deployment of edge AI-based algorithms. An important aspect in understanding the differences between the various platforms used for edge-based AI algorithm development and deployment is diving into their architecture. For this purpose, several existing edge AI platforms and workflows that allow integration with Arm® Cortex®-M4F-based MCUs have been benchmarked. While the best predictive accuracy can often be used to select the best-performing platform, comparing platforms for AI-based industrial applications can be a difficult task involving many architectural aspects. This paper provides an assessment and comparative analysis of some of the most essential architectural elements of differentiation (AEDs) in edge AI-based industrial applications, such as analytic capabilities in the time and frequency domains, features visualisation and exploration, microcontroller emulator and live tests, support for deep learning (DL), and using the ML core of the sensor. The use case selected for the benchmarking is a classification task based on the vibration of generic rotating equipment (e.g., motors), common to many industrial manufacturing applications. The benchmarking findings indicate that no single edge-AI-based platform can outperform all other platforms across all AEDs. The platforms have different implementation approaches and exhibit different capabilities and weaknesses. Nevertheless, they all produce independently relevant results, and together they provide an overall insight into their architecture and internal workings that can benefit the PdM solution. As they evolve and interact with each other, they will also overcome their weaknesses and gain strengths. Future work is intended to enlarge the comparison by considering additional edge AI platforms and AEDs.

Index Terms—Artificial intelligence, automated machine learning, edge computing, industrial automation, predictive maintenance, validation.

This work was conducted under the framework of the ECSEL AI4DI "Artificial Intelligence for Digitising Industry" project. The project has received funding from the ECSEL Joint Undertaking (JU) under grant agreement No 826060. The JU receives support from the European Union's Horizon 2020 research. (Corresponding author: O. Vermesan). O. Vermesan is with SINTEF Digital, Oslo, 0373 Norway (e-mail: Ovidiu.Vermesan@sintef.no). M. Coppola is with ST Microelectronics, Grenoble, 38019 France. (Marcello.Coppola@st.com).

I. INTRODUCTION

MAINTENANCE represents a significant portion of expenses associated with industrial manufacturing operations. Unexpected failures in production equipment/motors can cost significantly more than scheduled maintenance for an industrial manufacturing facility, and an unspecified amount of time can be required to fix problems. By using predictive maintenance (PdM) to continuously monitor equipment states to predict potential problems that may lead to costly failures, actions can be taken to prevent failures ahead of time. In PdM, the industrial manufacturing system is under constant monitoring, and analysis is made based on data collected from various sensors. As a result, the functioning of the equipment/motors can be optimised, and the costs of repair can be reduced.

The amount of data produced by industrial production processes has increased exponentially due to the rapid development of sensing technologies. When processed and analysed, sensor data can provide valuable information and knowledge about manufacturing processes, production systems and equipment. Across different industries, equipment maintenance plays an important role in a system's functioning and affects equipment operation time and efficiency. Hence, equipment faults need to be identified and solved to avoid shutdowns in production processes.

PdM involves leveraging sensor data to predict mechanical failures before they occur. It is a proactive maintenance technique that can predict the amount of time required to schedule maintenance activities for any system or any piece of equipment before it enters the failure mode of operation.

Industrial PdM is driven by physical models and processed data. The former employs physical laws to assess the degradation of the equipment/motors. The latter monitors various health indicators and employs methods such as ML and statistical approaches to find patterns in the data and determine operating conditions over time. There also exist combinations, such as rule-based methods, where ML or statistics are used to extract domain knowledge in the form of rules that govern model dynamics.

AI-based PdM refers to the ability of a PdM system to use knowledge and sensor data to identify, foresee and address potential issues before they lead to breakdowns in services, operations, processes, or systems. Different AI-based techniques can be explored for the implementation of industrial PdM systems: data-, ontology-, rule-, model-, sensor-, signal-, knowledge-, ML- and DL-based approaches. A survey of different AI approaches for PdM can be found in [12].

This article focuses on sensor-driven PdM using ML approaches with sensor data to predict failures over time, to minimise costs and to extend the useful life of components.

Sensor-driven PdM involves leveraging sensor data to predict failures before they occur. Rotating machine failures can be diagnosed and predicted by analysing the vibration signals derived from accelerometers connected to industrial equipment. However, sensor-driven PdMs are associated with many challenges. Transforming raw sensor data into actionable insights is a complex, time-consuming, and costly process requiring a systematic engineering approach to building, deploying, and monitoring ML solutions. Many aspects and questions need to be considered, such as the following:

1) How can distinguishable classes be defined?
2) What types of data will reveal the differences between classes?
3) What signal length will reveal the differences between classes?
4) What range of sensor values will fully reflect the range of input information?

The approach used in this paper includes elements of signal processing for analysing the equipment/motor's measured operating condition, identifying, and extracting the relevant PdM features and performing a diagnostic assessment based on prior knowledge of healthy systems.

The signal processing methods used include the time domain (standard deviation, trends, slope, and magnitudes), frequency domain (motor current signature analysis) and time frequency (Fourier transform, wavelet transforms, instantaneous power FFT, high resolution spectral analysis, wavelet analysis, bi spectrum, adaptive statistical time frequency method, etc.).

ML algorithms either classify the health state of the machinery or detect abnormal behaviour (e.g., any significant deviation from the normal operating condition). Bearings account for a large percentage of rotating machine failures, which can be predicted by analysing the vibration signals derived from accelerometers. The use case selected for benchmarking is a classification task based on accelerometer time-series raw data.

The paper is organised as follows. Section II provides background on edge AI processing and introduces the concepts of micro-, deep-, and meta-edge. The edge AI system architecture and the micro-edge acquisition approaches are introduced in Section III and IV. Section V gives an overview of the integration with edge AI platforms. In Sections VI–VII, the use case experiment, evaluation results, and further discussions are investigated, followed by conclusions in Section VIII.

II. EDGE AI PROCESSING

Recent advances in edge computing, edge AI and IIoT have contributed substantially to the deployment of lightweight PdM solutions at the edge to extend the lifetime of industrial equipment.

Edge computing is a paradigm where computation is executed on the edge of networks rather than on cloud servers, thus reducing the response time, transmission bandwidth, required storage, computation resources and network connectivity dependency. Sensor data can be processed in real-time, thus having the potential for PdM applications where fault diagnosis and dynamic control are time critical.

Edge AI increases the potential of PdM applications even further by merging AI/ML and edge computing, resulting in new algorithms for specific tasks that are less computationally expensive without compromising their effectiveness.

AI increases the value of IIoT by transforming data into useful information while IIoT increases the value of AI through connectivity and data exchange. Developments in intelligent applications and edge AI processing for industrial applications are reflected in advancements in different edge layers (micro-, deep-, meta-edge). The edge processing continuum includes the sensing, processing, and communication devices (micro-edge) close to the sensing/actuating elements, gateways, intelligent controllers processing devices (deep-edge), and on-premises multi-use computing devices (meta-edge). This continuum creates a multi-level structure

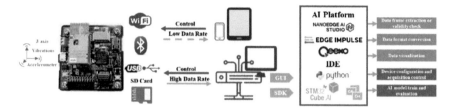

Fig. 1. Industrial edge AI system architecture.

that can advance processing, intelligence, and connectivity capabilities [11].

III. EDGE AI SYSTEM ARCHITECTURE

The overall PdM architecture proposed considers that the edge AI components are integrated into different edge layers.

The system implements an architecture integrated at the micro-, deep-, and meta-edge levels, allowing heterogeneous wireless sensor networks to communicate with the various gateways while integrating information from heterogeneous nodes in a shared on-premises edge server application and a shared database. The network architecture allows for interfacing with the existing SCADA system and providing a secure link to external cloud applications.

The micro-edge implementation increases information acquisition from the intelligent edge sensors placed on equipment/motors and allows end users to build predictive maintenance solutions based on advanced anomaly detection algorithms.

The heterogeneous architecture provides the ability to retrieve data from Bluetooth Low Energy (BLE), and Wi-Fi wireless sensor nodes using, for example, the MQTT protocol. This architecture has several advantages related to integrating data from heterogeneous sensor nodes, providing a mechanism for their transmission to an on-premises edge computing server, and creating geographically

distributed wireless sensor nodes over the production facility.

For this paper's use case, the edge AI components are deployed and evaluated at the micro-edge layer on the microcontroller and sensing units, while the edge AI platforms run at the meta-edge layer. The inference runs at the micro-edge on an Arm® Cortex®-M4F STM32L4R9ZIJ6 microcontroller and an ISM330DHCX MEMS sensor module containing ML capabilities.

IV. MICRO-EDGE DATA ACQUISITION APPROACH

The development of the industrial PdM is based on the design of sensors in combination with the Arm® Cortex®-M4F microcontroller. It incorporates all the required modules for data acquisition from the industrial equipment/motors, the pre-processing of the data, an interface for user interaction and a wireless network module for data transmission.

This micro-edge IIoT device (STWIN SensorTile Wireless Industrial Node) [3] used for the experiments comprises a 3-axis acceleration sensor (ISM330DHCX) and an Arm® Cortex®-M4F (STM32L4R9ZIJ6) 32-bit RISC core microcontroller. The device includes several other sensors and interfaces, but for the use case presented in this paper, the focus is mainly on these two components and the BLE, Wi-Fi, and serial interfaces as part of the

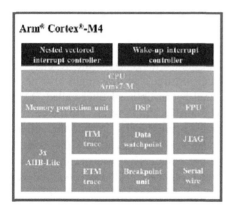

Fig. 2. STM32 Arm® Cortex®-M4F Microcontroller architecture.

communications used for implementing the use case.

The processing capabilities ensured by the microcontroller are 120 MHz, 640 KB SRAM, and 2 MB flash. The device's architecture is presented in Fig. 2.

The Cortex-M4 core implements a single-precision floating-point unit (FPU) that supports all the Arm® single-precision data-processing instructions and all data types. The core integrates a memory protection unit (MPU) which enhances the application's security and a set of digital signal processing (DSP) instructions. The microcontroller offers a fast 12-bit ADC (5 MSa/s), two operational amplifiers, two comparators, an internal voltage reference buffer, two DAC channels, a low-power RTC, two general-purpose 32-bit timer, 16-bit low-power timers, seven general-purpose 16-bit timers and two 16-bit PWM timers dedicated to motor control. For external sigma delta modulators (DFSDM), four digital filters are supported by the devices and up to 24 capacitive sensing channels are available. The device features standard and advanced communication interfaces such as: one DMA2D controller, three USARTs, two UARTs including one low-power UART, four I2Cs, three SPIs, two SAIs, one SDMMC, one CAN, one USB OTG full-speed, and one camera interface [9].

The interface between the micro-edge IIoT device and the deferent AI platforms is illustrated in Fig. 1. ISM330DHCX is a system-in-package (SiP) that works as a combo accelerometer-gyroscope sensor, generating acceleration and angular rate output data using a high-performance 3D digital accelerometer and 3D digital gyroscope. The ISM330DHCX features a full-scale acceleration range of $\pm 2/ \pm 4/ \pm 8/ \pm 16$ g full-scale range (FSR) and a different angular rate range of $\pm 125/ \pm 250/ \pm 500/ \pm 1000/ \pm 2000/ \pm 4000$ dps. The accelerometer's frequency range is from 1,6 to 6667 Hz and is selectable using specific output data rate (ODR).

The optimal ODR and FSR configuration depends on the use case. For example, using an ODR of over 1000 Hz will give practical detail when detecting and analysing motor vibrations. Using a larger FSR will allow for more considerable variations in sensor values, but the resolution will be compromised and have less detail.

The ISM330DHCX MEMS sensor module contains a Machine Learning Core (MLC), a Finite State Machine (FSM), and advanced digital functions that run custom algorithms; it shares the workload with the main processor to enable system functionality while saving considerable power and memory.

The dedicated MLC provides system flexibility, allowing some algorithms usually running in the dedicated microcontroller to be transferred to the MEMS sensor memory with the advantage of a consistent reduction in power consumption. MLC logic allows for the determination of whether a data pattern (for example, motion, pressure, temperature, magnetic data, etc.) matches a user-defined set of classes.

The ISM330DHCX MLC works on data patterns coming from accelerometers and gyro sensors, but it also allows for connecting to and processing data from external sensors (like magnetometer) using the Sensor Hub feature. The input data can be filtered using a dedicated configurable computation block containing filters and features computed in a fixed time window defined by the user. ML processing is based on logical processing containing a series of configurable nodes characterised by if-then-else conditions, where the feature values are assessed against specified thresholds.

ISM330DHCX's ML-processing capability originates from its decision tree logic. A decision tree is a mathematical tool comprised of a series of configurable nodes. Each node is described by an if-then-else condition, where an input signal (represented by statistical parameters computed from the sensor data) is assessed against a threshold.

The ISM330DHCX can be programmed to run up to 8 decision trees at the same time independently. The decision trees are kept in the device and provide results in the dedicated output registers. The decision tree's results can be read from the microcontroller at any time. Also, there is the option of providing an interrupt for every change in the results. The sensor data come from the 3-axis accelerometer.

The MLC inputs defined in the first block are used in the "Computation Block", where filters and features can be applied. The features are statistical parameters computed from the input data (or from the filtered data) in a defined time window, which the user may select. The features computed in the computation block are used as input for MLC's third block. This block, called "Decision Tree", includes the binary tree, which evaluates the statistical parameters computed from the input data. In the binary tree, the statistical parameters are compared against certain thresholds to generate the results. The decision tree's results might also be filtered by an optional filter called "Meta-classifier". MLC's results are the decision tree's results, which include the optional meta-classifier (i.e., a filter that uses internal counters to evaluate the decision tree's outputs) [10]. Decision tree generation for ISM330DHCX in IIoT edge devices can be achieved using a dedicated tool available as an interface extension (Unico GUI) or using external tools, such as MATLAB, Python, RapidMiner, Weka.

V. INTEGRATION WITH EDGE AI PLATFORMS

Deploying AI on the edge in the view of PdM is gaining increasing interest in various industries, but this deployment is prone to challenges in determining the optimal solution architecture for a given application. To a certain extent, these challenges are due to a lack of robust frameworks that can integrate multivariate data from heterogeneous sensors, partly because validation of the design is difficult, which means the predictions derived from these sensors' data may not be trustworthy.

PdM solution architecture is the foundation for industrial applications because it helps in adapting IT implementations to specific business needs and describing their functional and non-functional requirements and implementation stages. It comprises many components and processes that draw guidance from various architectural viewpoints.

Recently, new edge intelligence platforms have emerged that may be good candidates for a PdM solution architecture. These AI platforms must be able to leverage sensor data, build AI solutions for constrained environments and deploy on edge devices. However, they use various components – including data gathering, model training and inference implementation – to implement core tasks, and they include various degrees of automation, control, and transparency into their processes.

Thus, investigating the architecture of edge AI platforms is an important step in understanding the differences among them. The approach adopted in this work for benchmarking was to define a set of architectural elements that make the differences among the essential platforms explicit. These elements affect theseplatforms' performance (accuracy and loss) and other relevant qualities.

In this paper, three existing edge AI platforms for integrating AI mechanisms within MCUs have been employed:

1) Qeexo AutoML (Qeexo) - automated ML platform for Arm® Cortex®-M0-to-M4-class processors,
2) NanoEdge^{TM} AI (NEAI) Studio,
3) Edge Impulse (EI)

The edge AI platforms are used to deploy ML algorithms and run pre-trained Artificial Neural Networks (ANN) with on-chip self-training on an ultra-low power Arm® Cortex®-M4F STM32L4R9ZIJ6 microcontroller running at 120 MHz and the MLC of the ISM330DHCX MEMS sensor module.

VI. MAINTENANCE CLASSIFICATION CASE STUDY

In this paper, classification has been used as a case study. The following steps were used for implementation:

1) **Data acquisition** includes the acquisition sensor setup for each specific edge AI platform, data retrieval over wired or wireless connectivity, data labelling and storage of vibration data in a specific format.
2) **Condition monitoring** includes data cleaning/denoising, data visualisation, pre-processing, feature extraction and feature engineering.
3) **Anomaly detection and classification** include ML of the system's behaviour, unsupervised learning at the edge for anomaly detection and supervised learning to classify states.
4) **Model emulation and model deployment** at the inference on the target device incorporate the remaining life prediction models, overall efficiency optimisation and operational system integration.

A. Use Case Design

The classes were defined based on conditions (motor speeds) and sub-conditions (malfunctions). The motor was operating at fixed speeds, minimum, medium and maximum. A malfunction of the motor (motor fan trepidations) was added on all classes to obtain new ones. The classes defined are:

1) Class A: the motor is running at minimum speed
2) Class B: the motor is running at half of the speed
3) Class C: motor is running at maximum speed
4) Class D: the motor is running at minimum speed with an excess load producing a centrifugal force
5) Class E: the motor is running at half of the speed with an excess load producing a centrifugal force
6) Class F: the motor is running at maximum speed with an excess load producing a centrifugal force

The classes were defined with the following goals in mind:

1) The motor behaviour and the classification problem being solved with ML/DL were studied in-depth.
2) Classes should be distinguishable for easier classification.
3) Data sets should be class-balanced.
4) Data sets should be properly split (training, validation, test).

B. Experimental Setup

The design and implementation steps and the experimental setup of the end-to-end (E2E) classification application consist of three primary workflows, including NEAI Studio, EI and Qeexo platforms.

All three workflows are E2E development platforms for embedded AI/ML, supporting developers through all phases of their projects, including collecting, pre-processing, and leveraging sensor data; generating and training the models; and deploying to highly constrained environments.

The experimental process occurred as follows: vibration signals for each class were collected live from the micro-edge IIoT device mounted on the motor, using NEAI and Qeexo, as EI currently does not support direct connection to the STWIN Sensor Tile. The recorded signals were then analysed in both the time and frequency domains, and then filtrated, exported, and converted to prepare the cross-platform datasets. Several AI models were then built with each of the three flows, using the accelerometer spectral features, and optimised for performance and resource constraints. In the end, the three models were deployed for MCUs, and classification inferences were run to note real-time performance. During all stages, comparative analysis has been performed, both in terms of the processes involved and the results produced.

All components and processes across the E2E workflow can benefit from automation, parameter control, transparency, integration, and other facilities. For instance, automation removes tedious, iterative, and time-consuming work, thus making AI/ML more accessible to applications. Nevertheless, when embedding AI/ML at the edge, some of these components and processes present particularities, and therefore, have been selected as elements of differentiation in the comparative analysis of the three platforms. These are:

1) Data acquisition methodology
2) Pre-analysis in time and frequency domain
3) Visualization and exploration of features
4) Support for various ML algorithms and ANN
5) Emulation/live test capabilities
6) Sensor MLC capabilities
7) Deployment automation

C. Data Acquisition Methodology

Data acquisition parameters, such as buffer size, signal length and sampling frequency, vary depending on project constraints. In general, the higher the sampling frequency, the higher the chances of capturing important features in the signal snapshot; however, higher sampling affects the constraints, such as memory and power consumption.

The motor's vibration patterns were analysed at different frequencies, and 1667 Hz was identified as the most suitable sampling frequency to capture the patterns.

As the task was to compare the platforms, the goals were to apply configurations that were as similar as possible and to ensure that the same signals were used for processing, thereby yielding comparable results. This proved to be a difficult task, because the sampling frequency, buffer size and signal length are inter-related, and their interrelationships also vary, depending on what each platform defines as fixed or configurable. Thus, the sampling method is a differentiating element.

In the case of NEAI, the signal length was determined based on buffer size and sampling frequency, namely approximately 300 milliseconds (= 512/1667) for a buffer size of 512 samples on each axis, in total 1536 values per signal.

In the case of Qeexo, the length of the signal exported is fixed, 50 milliseconds, and the buffer size is approximately 83 samples: 50/(1000/1667). The buffer size can also vary, due to sample rate tolerance.

Another differentiating element is the duration of a live collection session, which, for practical reasons, was defined as being about 30 seconds. With Qeexo, the collection duration was easily set to the exact number of seconds, while with NEAI, the duration was determined by the number of signals per collection (in this case, set at 100 to obtain sessions that were 30 seconds long).

All three platforms enabled the acquisition of data by both directly collecting them from sensors and uploading them from files. The acquisition methodology consisted of three steps:

1) Collect sensor data using platforms that enable direct connection to the microcontroller (e.g., NEAI and Qeexo).
2) Export the recorded data and convert it to a cross-platform format.
3) Generate cross-platform data sets (split for training, validation, and test) and then upload.

The acquisition methodology is depicted in Fig. 3. The data recorded and exported using NEAI had to be converted into formats accepted by EI and Qeexo.

Fig. 3. Cross-platform conversion of sensor data (collected vs uploaded).

Similarly, the data recorded and exported using Qeexo had to be converted into formats accepted by NEAI and EI. EI acted as a neutral platform, processing no data collected on its own, only data from the other two platforms.

Regarding the collection process, measurements could be collected directly from the sensor IoT devices within their GUIs, but the processes differed. In the case of NEAI, acquiring signals was straightforward when using the datalogger functionality, requiring only an SD card. For the experimental use case, a simple logger application was built in C to read the raw accelerometer sensor data and log it directly onto the serial port.

In the case of Qeexo, a data collection application was deployed to MCU with the press of a button, after the ODR(Hz) and FSR(g) parameters have been configured.

D. Pre-analysis in Time and Frequency Domain

All three platforms have pre-analysis and pre-processing capabilities in both the time domain and the frequency domain. This is shown in Fig. 4, Fig. 5 and Fig. 6, which visualise the frequency plots for the F class along the accelerometer z-axis.

In the case of NEAI, the graph's x-axis in the temporal plots corresponds to the number of samples (512), while the y-values represent the mean value of each sample across all 100 signals, their min-max values, and their standard deviation.

In the frequency domain, the x/y axis shows frequency/amplitude. With Qeexo, the frequency domain is a frequency spectrogram.

In the frequency domain, it is often easier to differentiate the individual classes, thus ensuring a high accuracy score. In NEAI, switching to the frequency domain is done by activating the filter settings in the signals pre-processing step. This allows for setting the filter frequency parameters, so that only the frequencies that represent the characteristics of the motor vibration are kept and the rest are attenuated. Filtering techniques also help eliminate both the high frequency noise that interferes with the vibration signal and the frequencies for transitions between states, which would normally yield an unknown class.

In EI, the Spectral Analysis processing block extracts the frequency and power characteristics of a signal over time. This component must be included

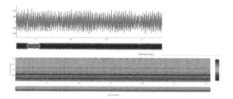

Fig. 6. F class along the y-axis (Qeexo) in time and spectrogram.

explicitly in the design, otherwise the raw accelerometer data will be used without pre-processing.

E. Visualisation and Exploration of Features

Standard classification algorithms are not well suited to work on time series; therefore, raw time series data is sampled using various windowing techniques and aggregated to generate new features, such as mean, standard deviation, RMS, median, number of peaks, skewness, kurtosis, and energy.

If the data exhibit clustering, retaining only the features with the highest variance can make the clusters more visible, with clear distinctions between them (i.e., the clusters do not overlap, and some space exists between them). Making the clusters more visible can be obtained by applying dimensionality reduction and visualisation techniques, such as principal component analysis (PCA), t-distributed stochastic neighbourhood embedding (t-SNE) and uniform manifold approximation and projection (UMAP).

While PCA preserves the data's global structure at the expense of the local structure (which might get lost), t-SNE aims to embed the points from higher dimensions to lower dimensions by fitting the data into a distribution such that the neighbourhood of the points is preserved. UMAP provides a balance between local and global structures. Both t-SNE and UMAP first build a graph

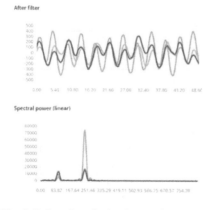

Fig. 4. F class along the 3-axis (NEAI)

After filter

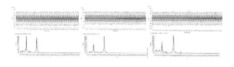

Spectral power (linear)

Fig. 5. F class along the 3-axis over 50 ms window (EI)

that represents data in high dimensional space, then reconstruct the graph in a lower dimensional space while retaining the structure.

PCA is employed by Qeexo, while UMAP is employed by both Qeexo and EI. EI also generates a list of features sorted by importance, indicating how vital each feature is for each class compared to all other classes. RMS, power density and peak values of vibration along the 3-axis proved to be the most important features for determining the class. Based on this information, the less critical features can be removed from the training set to simplify the model and make it more manageable while maintaining its relevance and performance.

Most AI-based platforms have embedded visualisation and feature exploring functionality in the form of tables or graphs, where dimension reduction algorithms are employed to generate projections from the higher-dimensional feature space onto two dimensions. The features are coloured based on labels; thus, it is easy to observe if distinct labels are separated based on the available features.

The generated two-dimensional plots proved to be a good indicator for how well the motor classifier will perform. The fact that the features were visually clustered was a good indication that the model could be trained to perform the classification with high accuracy.

When the features overlapped to a large degree, as was the case in the early development stages, it was difficult for the trained models to distinguish the classes. The problem was addressed by collecting more signals and increasing the size of the sampling signal to better capture the signal patterns; sometimes, it was necessary to redefine the classes.

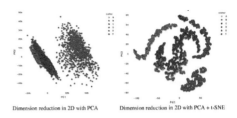

Fig. 7. Dimension reduction with two components (PC1 and PC2) with PCA (left) and PCA+t-SNE (right)

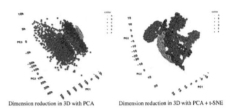

Fig. 8. Dimension reduction with three components (PC1, PC2 and PC3) with PCA (left) and PCA+t-SNE (right)

Notably, the lack of separability of the classes in the two-dimensional plots generated by the platforms does not necessarily imply a lack of separability between classes in the higher-dimensional space. This matter was further explored off-platform (Python framework), with various features and dimension reduction techniques along both 2D and 3D plots.

As shown in the scatter plot in Fig. 7, PCA with two components did not provide sufficient insights into the different classes. On the other hand, t-SNE is known for its ability to capture non-linear dependencies, thus combining the two provides better results. In general, applying PCA in conjunction with either t-SNE or UMAP provides an initial reduction in the dataset's dimensionality, while still preserving most of the important data structure.

As shown in Fig. 8, applying dimensionality reduction with the three components provided even better results.

F. Support for Various ML/DL Algorithms and Automation

All three platforms offer an automated mechanism for generating the AI model architecture and training, but its implementation differs considerably, mainly due to the type of algorithms employed.

Fig. 9. Benchmarking with NEAI. All correctly classified (green dots)

NEAI employs several ML models implementations such as K-Means, Random Forest (RF), Support Vector Machine (SVM), optimised for embedded systems, each having their own hyper-parameters. The benchmarking of the classification task involved searching through these algorithms and producing combinations of three elements: pre-processing, ML algorithm, and hyper-parameters. These combinations, called libraries, are displayed as a ranked list that is evaluated based on accuracy, confidence, and memory usage, with the highest accuracy being on top. Accuracy denotes the library's ability to correctly classify each signal into the right class, whereas confidence reflects the library's ability to distinguish between classes. Learning is fixed at library generation based on the data provided for each class. Selecting the suitable model occurs after training.

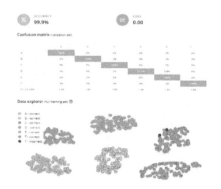

Fig. 10. Benchmarking with EI. Confusion matrix and data explorer. Correctly classified (green dots) and misclassified (red dots).

In contrast to NEAI, EI and Qeexo allow explicit selection and configuration of the algorithms of choice before training. EI supports ML algorithms as well as NN architectures following the full pipeline that is commonly found in traditional deep learning frameworks, such as Keras and TensorFlow, thereby making it more flexible.

EI employs K-means algorithms for anomaly detection, and Keras for classification and regression tasks.

Qeexo also supports ML algorithms, such as Gradient Boosting Machine (GBM), XGBoost (XGB), RF, Logistic Regression (LR), Decision Trees (DT), Gaussian Naive Bayes (GNB) as well as Artificial Neural Network (ANN), Convolutional Neural Network (CNN), Recurrent Neural Network (RNN) architectures, but is fully automated, and the selected and trained model can be deployed to target embedded hardware with just one click, without any coding required. Solutions are optimised to have ultra-low latency, power consumption and a small memory footprint.

The benchmarking allowed for comparison of the performance of different model architectures generated by the same platform, but also of the performance of the same model across platforms. The confusion matrix has been a useful tool.

Snapshots from the benchmarking with NEAI, EI and Qeexo are presented in Fig. 9, Fig. 10, Fig. 11 respectively, showing comparable results on the validation data.

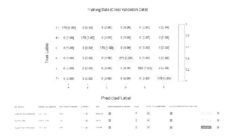

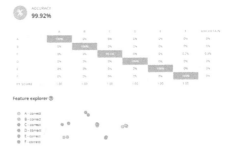

Fig. 12. Live streaming evaluation of SVM trained model using NEAI Emulator (left) and trained ANN model Qeexo (right).

Fig. 11. Benchmarking with Qeexo. Confusion matrix for SVM model (up). Overview trained models (down).

G. Emulation, Live Test Capabilities

Testing the trained model performance using the testing dataset was done to analyse how well the model performs against unseen data prior to deployment on the target MCU. To ensure unbiased evaluation of the model's effectiveness, the test samples were not used directly or indirectly during training.

Fig. 13. EI ANN model testing with test datasets (Arm® Cortex®-M4 MCU STM32L4R9 not yet supported).

Testing was also performed live, while changing motor speeds and triggering shaft disturbances, to switch between classes and cover all six classes. Live testing ensured unbiased evaluation of the model's effectiveness with completely new signals, not seen before.

The results of the tests are depicted in Figures 12, 13, and 14 showing that all three classifier systems properly reproduced and detected all classes with reasonable certainty percentages, and these are comparable. This was the case both when testing on test datasets and when testing live, although the latter had some particularities. When testing live with NEAI, the trained model runs on a microcontroller emulator. In the case of Qeexo, live means both new data and the trained model running in the microcontroller.

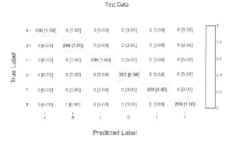

Fig. 14. Qeexo testing with test datasets collected (SVM model).

algorithms and features. Experiments revealed that the only available algorithm (e.g., decision tree) performed well on the collected data, demonstrating sufficient predictive value to distinguish the classes with good accuracy.

H. Sensor MLC Capabilities

The MLC capabilities were only explored with Qeexo. Deploying in the MLC implies limited availability of the

I. Deployment Automation

Model deployment was the final stage in the E2E workflow, consisting of flashing the compiled binary into the MCU, running on-device inferences, and displaying the expected output. All three workflows were successfully deployed, and the models were able to accurately recognize motor states in real time.

However, the deployment steps exhibited some particularities with the three platforms. Because deployment occurred in the context of micro-edge embedded systems, the steps depended on the hardware and software used.

In the case of NEAI, the trained model was deployed in the form of a static library (libneai.a), an AI header file (NanoEdgeAI.h) containing functions and variable definitions, and a knowledge header file (knowledge.h) containing the model's knowledge.

For the EI deployment, the CMSIS-PACK for STM32 packaged all signal processing blocks, configuration and learning blocks up into a single library (.pack file), which was then added to the STM32 project using the CubeMX packages manager.

In the case of Qeexo, the trained model was deployed to MCU with the press of a button, without any coding required. The model was compiled, built, and flashed onto the MCU target automatically.

All three platforms generated an optimised code for inclusion in the microcontroller application. However, while the integration was accomplished within the integrated development environment (IDE) for NEAI and EI, for Qeexo, it was done automatically with the press of a button in the GUI.

This led to the conclusion that the three platforms could co-exist in the development portfolio of a PdM solution provider and be employed depending on the application or stage of development. NEAI could be used for rapid prototyping, as it enables automatic exploration using various traditional ML algorithms and emulation to test performance without actual deployment to the target MCU. Qeexo could be used to automate repetitive deployments to the target MCU. Finally, EI could be used for specialised NN architectures in which more control over the model's parameters and hyperparameters is desired (using Keras).

VII. RESULTS AND DISCUSSION

Although automation compensates for many of the drawbacks of manual processes, it is important to verify that the E2E workflow is easily repeatable and reproducible, i.e., to validate the design.

One of the main findings during experimentation is that in addition to the primary design flow with the AI platform of choice, at least one complementary design flow with another AI platform is necessary for validation. This complementary flow will have a similar purpose during development as the parallel flow has in operation; namely, to avoid a single point of failure.

For instance, employing the same ML algorithm in both the primary and complementary flows would hardly produce the same results, as the implementations are different. However, if they produce similar results, it will increase the level of confidence in the design. Extremely different results will provide valuable insight into what validation actions are needed. In this way, the complementary flow may compensate for eventual flaws in the primary flow.

VIII. CONCLUSION

This paper presented three design workflows with different edge AI platforms and embedded inference engines used for the same classification use case, highlighting the different aspects of model design, development, and deployment of AI-based industrial applications approached by edge AI multi-platform solutions as part of a holistic flow framework for industrial PdM applications.

The three platforms are Qeexo AutoML, NanoEdge™ AI Studio and Edge Impulse for integrating edge AI mechanisms at the micro-edge within MCUs such as Arm® Cortex®-M4F STM32L4R9ZIJ6 microcontroller and ISM330DHCX MEMS sensor module containing an MLC, an FSM, and advanced digital functions that are used to run custom algorithms on the inertial measurement unit (IMU) and share the workload from the central processor enabling system functionality while significantly saving power and memory footprint.

Each platform was benchmarked by assessing some of the most critical AEDs using different ML algorithms and NN implementations for an industrial PdM application to classify the data from motor vibrations measured with a 3-axis accelerometer IIoT device. The results were benchmarked by considering specific edge AI flow frameworks to analyse the results.

Transforming raw sensor data into actionable insights is complex, time-consuming, and costly, requiring a systematic engineering approach to building, deploying, and monitoring ML solutions.

The benchmarking findings indicate that no single edge-AI-based platform can outperform all other platforms across all AEDs. The platforms have different implementation approaches and exhibit different capabilities and weaknesses. Nevertheless, they all produce independently relevant results, and together they provide overall insight into their architecture and internal workings that can benefit the PdM solution. As they evolve and interact with each other, they will also overcome their weaknesses and gain strengths. Future work is intended to enlarge the comparison by considering additional edge AI platforms and AEDs.

REFERENCES

[1] J. Lin, W.-M. Chen, Y. Lin, C. Gan, S. Han et al., "Mcunet: Tiny deep learning on IoT devices," Advances in Neural Information Processing Systems, vol. 33, pp. 11 711-11 722, 2020. Available:https://procee dings.neurips.cc/paper/2020/file/86 c51678350f656dcc7f490a43946ee5 -Paper.pdf

[2] J. Lin, W.-M. Chen, H. Cai, C. Gan, and S. Han, "Mcunetv2: Memory efficient patch-based inference for tiny deep learning," in Annual Conference on Neural Information Processing Systems (NeurIPS), 2021. Available: https://arxiv.org/pdf/21 10.15352.pdf

[3] STMicroelectronics, 2022, STWIN SensorTile Wireless Industrial Node development kit and reference design for industrial IoT applications. Available: https:r//www.st.com/en/e valuation-tools/steval-stwinkt1b.h tml

[4] STMicroelectronics, 2022, X-Cube-AI - AI expansion pack for STM32-CubeMX. Available: https://www.st .com/en/embedded-software/x-cube -ai.html

[5] STMicroelectronics, 2022, Nano-Edge AI Studio V3. Available: ht tps://stm32ai.st.com/

[6] Edge Impulse. Edge Impulse Development Platform. Available: https: //www.edgeimpulse.com/

[7] Qeexo, Qeexo AutoML Platform. Available: https://qeexo.com/

[8] STMicroelectronics, 2022, Nano-Edge AI Studio V3. Available: ht tps://stm32ai.st.com/

[9] STMicroelectronics, 2022, STM32-L4R9ZI. Available: https://www.st .com/en/microcontrollers-micropr ocessors/stm32l4r9zi.html

[10] STMicroelectronics, 2022, ISM-330DHCX. Available: https://www.st.com/en/mems-and-sensors/ism330dhcx.html

[11] O. Vermesan and M. Coppola, "Embedded Edge Intelligent Processing for End-To-End Predictive Maintenance in Industrial Applications," in Industrial Artificial Intelligence Technologies and Applications, O. Vermesan, F. Wotawa, M. Diaz Nava end B. Debaillie, eds. Gistrup, Denmark: River Publishers, 2022, Ch. 12, pp. 157-175. Available: https://www.riverpublishers.com/pdf/ebook/chapter/RP_9788770227902C12.pdf

[12] Y. Ran, X. Zhou, P. Lin, Y. Wen, and R. Deng, "A Survey of Predictive Maintenance: Systems, Purposes and Approaches". IEEE Communications Surveys & Tutorials, 20, pp. 1-36. https://arxiv.org/pdf/1912.07383.pdf

Ovidiu Vermesan is Chief Scientist at SINTEF Digital, Oslo, Norway, where he is involved in applied research on edge AI and future edge autonomous intelligent systems, smart systems integration, wireless sensing devices and networks, microelectronics design of integrated systems (analogue and mixed-signal), IIoT.He holds a PhD in Microelectronics and a Master of International Business (MIB).His applied research activities focus on wireless and smart sensing technologies, advancing edge AI processing, wireless and smart sensing technologies, embedded electronics, and the convergence of these technologies and applying the developments to applications such as autonomous systems, green mobility, energy, buildings, electric connected, autonomous, and shared (ECAS) vehicles,

and industrial manufacturing. He is currently the technical co-coordinator of the ECSEL JU "Artificial Intelligence for Digitising Industry" (AI4DI) project.

Marcello Coppola is technical Director at STMicroelectronics. He has more than 25 years of industry experience with an extended network within the research community and major funding agencies with the primary focus on the development of breakthrough technologies. He is a technology innovator, with the ability to accurately predict technology trends. He is involved in many European research projects targeting Industrial IoT and IoT, cyber physical systems, Smart Agriculture, AI, Low power, Security, 5G, and design technologies for Multicore and Many-core System-on-Chip, with particular emphasis to architecture and network-on-chip. He has published more than 50 scientific publications and holds over 26 issued patents. He authored chapters in 12 edited print books, and he is one of the main authors of "Design of Cost-Efficient Interconnect Processing Units: Spidergon STNoC" book. Until 2018, he was part of IEEE Computing Now Journal Technical editorial board. He contributed to the security chapter of the Strategic Research Agenda (SRA) to set the scene on R&D on Embedded Intelligent Systems in Europe. He has served in different roles at numerous top international conferences and workshops. He graduated in Computer Science from the University of Pisa, Italy in 1992.

Food Ingredients Recognition Through Multi-label Learning

Rameez Ismail and Zhaorui Yuan

Abstract—The ability to recognize various food-items in a generic food plate is a key determinant for an automated diet assessment system. This study motivates the need for automated diet assessment and proposes a framework to achieve this. Within this framework, we focus on one of the core functionalities to visually recognize various ingredients. To this end, we employed a deep multi-label learning approach and evaluated several state-of-the-art neural networks for their ability to detect an arbitrary number of ingredients in a dish image. The models evaluated in this work follow a definite meta-structure, consisting of an encoder and a decoder component. Two distinct decoding schemes, one based on global average pooling and the other on attention mechanism, are evaluated and benchmarked. Whereas for encoding, several well-known architectures, including DenseNet, EfficientNet, MobileNet, Inception and Xception, were employed. We present promising preliminary results for deep learning-based ingredients detection, using a challenging dataset, Nutrition5K, and establish a strong baseline for future explorations.

Index Terms—Automated diet assessment, deep learning, visual ingredients recognition, machine learning, multi-label learning.

Rameez Ismail and Zhaorui Yuan are with the team of embedded intelligence and analytics at Philips Research, High Tech Campus, 5656, Eindhoven, The Netherlands, Corresponding author: R. Ismail (rameez.ismail@philips.com). The research leading to these results has received funding from the European Union's ECSEL Joint Undertaking under grant agreement *n°* 826655 - project TEMPO.

I. INTRODUCTION

WHAT we eat and drink has a huge impact on our daily lives and our wellbeing. It is well established by now that a healthy and well-balanced diet is paramount to one's health. Daily diet varies considerably around the world, however, people in almost all regions of the world could benefit from rebalancing their diets by eating optimal amounts of various nutrients [1]. A suboptimal diet does not only carry risks for physical health but might reduce cognitive capabilities. Understandably, numerous studies, for example [2][3][4], implicate dietary factors in the cause and prevention of diseases such as, cancer, coronary heart disease, diabetes, birth defects, and cataracts. Similarly, findings from nutritional psychiatry indicate multitude of consequences and implications between what we eat and how we feel and ultimately behave [5]. On a population scale, eating habits and a broader diet pattern of a population is shown to be correlated with its health outcomes and longevity [6][7]. For example, a diet that adheres to traditional Mediterranean diet principles represents a healthy pattern and is positively associated with the longevity in Mediterranean blue-zones [8]. Although, such guidelines and general principles are quite useful, the individual metabolic responses varies substantially even if individuals are eating identical meals

[9]. This calls for a personalized nutrition guidance approach that goes beyond general health recommendations. Personalized or precision guidance could additionally enable better management of various nutrient-related health conditions and diseases [10], for example, celiac disease, bowel syndrome, phenylketonuria, food allergies and diabetes. However, to make the proposition viable and to drive impact at scale, the guidance system must be automated.

A major difficulty in realizing an automated diet guidance is to capture the eating habits of a user accurately and effortlessly. Besides providing an introspection ability to the consumers, capturing dietary data from several participants is fundamental for understanding the diet and disease relationship. An accurate assessment of dietary intake enables the investigators to make progress in diet related studies by discovering patterns in context of nutritional epidemiology [11][12]. Diet assessment is usually performed using one of the three basic methods: meal recall, food diaries, or food frequency questionnaires. However, all these methods are based on self-reporting and therefore are time consuming, tiresome, and prone to misreporting errors. With recent advances in sensor technology and Machine Learning (ML) algorithms, automated food assessment has gained ground. Some of the technologies being explored in this direction include digital biomarkers and indigestible sensors [13][14], imaging sensors coupled with advanced AI/ML algorithms [15][16], as well as smart scales and eating utensils [17]. Digital cameras are readily available and can perform a detailed analysis on the food being consumed and the eating behaviors of the subjects but relies heavily on the visual recognition and the interpretation technology.

This work focusses on evaluating the performance of various state-of-the-art ML algorithms to detect ingredients or food-items present in a digital food image. Recognition of the actual ingredients is one of the two essential ML functions towards our envisioned food assessment system depicted in Fig. 1. The other is the portion estimation. When all the major ingredients and their portion size are identified, a descriptive nutrition log can be created by using a simple lookup service that searches across various nutrition fact databases. There are several food databases in public domain which can be utilized for this purpose, such as USDA Food and Nutrient Database for Dietary Studies (FNDDS) and Dutch Food Composition Database (NEVO).

In contrast to building separate ingredient recognition and portion estimation functions, some studies attempted an end-to-end nutrition estimation scheme [18], which estimates the total amount of macro-nutrients directly from the

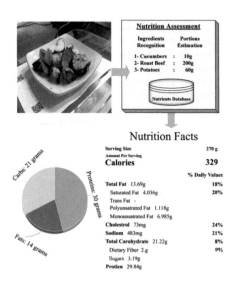

Fig. 1. The envisioned approach towards automated nutrition assessment. The assessment block consists of two ML functions, one for detecting the ingredients and the other to estimate the quantity, and a nutrients database look-up service.

food images. Such end-to-end schemes, however, suffer from descriptive inadequacy. For example: actual food items composition and the respective portions remain undiscovered, which are important details for dietary research and for determining the eating habits of the users. Additionally, as the two-stage scheme depicted in Fig. 1 is more transparent, it is also expected to score better on the scale of user engagement and trustworthiness.

In summary, automated nutrition assessment has the potential to empower communities, on one hand, while enables investigators to make progress in nutritional epidemiology on the other hand. The main contributions of this work are: (a) motivate and investigate the problem of multi-item food recognition and proposes a multi-label learning framework to achieve this, and (b) evaluates and benchmark the proposed framework using various state-of-the-art deep learning modules and a challenging dataset, Nutrition5k [18], comprised of real-world food images with ingredient level annotations.

II. RELATED WORK

Automated food assessment through visual recognition has attracted a decent research interest in recent years. Existing efforts include deep learning based single dish recognition [19][20], contextualized food recognition (for example using GPS data which exploits the knowledge about location and data from the restaurants), multiple-food items detection [21][22] and real-time recognition [23]. This section mainly reviews the previous works on multiple food-item recognition using dish images as well as multi-label learning in general.

Some of the early approaches, towards multiple food-item recognition, relied on a separate candidate regions generation, using either simple circles or De-

formable Parts Models (DPM), followed by features based classification of the candidate regions [24]. Such approaches employed various hand-crafted features such as color, texture, and Scale-invariant Feature Transform (SIFT) features for classifying the candidate regions. However, majority of the recent work is based on deep neural networks, because of its powerful feature representation ability, where such features are directly learnt from the data. For example, [25] proposed to use a convolution neural network GoogLeNet for labels prediction and employed the DeepLab [26] for semantic segmentation of the images. This pixel level image segmentation allows further analysis, such as estimation of the count and portion of the food constituents. More recently, PRENet [27] adopted a progressive training strategy to learn multi-scale features for a large-scale visual food recognition task. The approach utilizes a self-attention mechanism to contextualize the local features. These refined local features and a set of global features are then concatenated for the final classification task.

Both pure convolutional [20][25][18] and attention-based networks [27] have shown promising results for visual food recognition. Nevertheless, the latter can contextualize and dynamically prioritize the information. This suggests that the attention networks can extract much richer descriptions from the images compared to pure convolutional networks. However, it also implies that the learning process is comparatively difficult as the attention-based networks have more degrees of freedom. Moreover, since attention mechanism is a novel paradigm within deep learning, its potential is not yet fully explored for the task of ingredients and multiple food-item recognition.

Dish images in real-life usually contain multiple food items and ingredients, which makes it worthwhile to detect multiple labels independently for each input image. It also provides an added benefit that any food image can be inferred by the model, even if the actual dish is novel to the model, given that all its constituents were taught to the network during the training phase. The independent prediction of various labels against a single image is a more general classification problem, commonly known as multi-label classification or extreme classification. In such a regime of classification [28][29], the goal is to predict the existence of a multitude of classes, thus forcing the model and training scheme to be efficient and scalable. The networks usually contain a stem and a head: the stem outputs a spatial embedding while the head transforms this spatial embedding into prediction logits. The most employed head is a Global Average Pooling (GAP) layer, which computes a scalar global embedding for each spatial embedding, followed by a fully connected read-out layer.

In case of multi-label learning, each neuron in the read-out layer is a binary classifier representing a specific class. Recently, several works proposed novel attention-based heads for multi-label classification. For example, [30] proposed an approach which leverages Transformer [31] decoders to query the existence of various class labels. Similarly, [32] introduced a class specific residual attention (CSRA) module that generates class-specific features, using a simple spatial attention score, and combines them with the global average pooling achieving state-of-the-art results. ML-Decoder [33] is yet another classification head based on transformer decoder architecture, which outperforms on various multi-label classification. One of the critical challenges for multilabel image

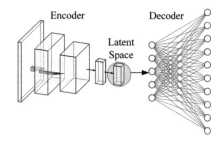

Fig. 2. The meta-structure used for the evaluated models.

classification is to learn the inter-label relationships and the dependencies.

The transformer-based methods employ attention mechanism with the goal to implicitly model the co-occurrence probability of the labels through relative weighing of the latent embeddings. Correspondingly, graph convolutional networks for multilabel learning [34] is another emerging theme that captures the complex inter-label associations by modelling them as graph nodes. In this work, we constrained our analysis to networks that follows a definite meta-structure shown in Fig. 2. The encoder block extracts the discriminative features from the image. Thus, projecting the image onto a latent space, while the decoder predicts the presence of various labels using the latent feature space.

III. METHODOLOGY

In this section, we first describe various deep learning modules used to build the classification networks benchmarked in this work. Then we explore the dataset, *nutrition5k,* used to train and evaluate the networks, followed by explanation of the loss function, utilized for the training, and the evaluation metrics.

A. Multi-Label Classification Networks

All classification networks benchmarked in this work are composed of an encoder and a decoder module, as depicted in Fig. 2. The encoder performs features extraction, using the input image, while the decoder exploits these features to construct an ingredient breakdown sequence. The decoder can be as simple as a linear learnable projection that maps the latent space on to a layer of read-out neurons but can also have a complex construction. For example, a versatile attention block that exploits inter-label relationships through a querying mechanism or an autoregressive decoder that contextualizes its mapping based on its previous predictions. Next, we describe the various implementation of these blocks implemented and benchmarked in this work:

1) **Encoder**

An encoder block can be modelled by any network that projects the image onto a latent space, $F \in \mathbb{R}^{H \times W \times D}$. For example, both convolutional neural networks, which progressively construct richer representations using convolution operations, and transformer networks, which transforms images into a small set of embedding tokens, are great candidates for the encoder block. In this work, however, we limited our analysis to convolutional networks for encoding. The employed encoding networks include, MobileNet [35][36], DenseNet [37], Inception network [38], Xception network [40] and EfficientNet [39]. These are some of the top-performing networks, when tasked to perform single label classification, evaluated on a large-scale visual recognition challenge ImageNet [41] dataset.

2) **Decoder**

We used two distinct approaches to model the decoder block, the first approach employs a simple GAP based decoder while the second exploits an attention-based group-decoding scheme, ML-Decoder [33]. In a GAP based decoder, we first project the features, $F \in \mathbb{R}^{H \times W \times D}$, on to a single dimensional vector, $z \in \mathbb{R}^{D}$, by averaging over the spatial dimensions. Afterwards, a dense layer transforms the single dimensional vectors through a learnable linear projection into K logits. K is the number of total classes. The ML-Decoder is adapted from the transformer-decoder [31] with the goal to meet the computational demands for multi-label learning as the computational cost grows quadratically with the number of classes. The proposed modifications include the removal of a self-attention block, which reduces the quadratic dependency of the decoder in the number of query tokens to a linear one, and the introduction of a group decoding scheme. In a group-decoding scheme, a single query token is responsible for decoding multiple ingredients, thus limiting the required number of query tokens.

This strict decoupling of the roles enables the reuse of the latent space across various image recognition tasks. Fig. 3 provides a visual description of the two decoding schemes evaluated for the task of multi-label learning.

B. Dataset

We employed Nutrition5k [18] to train and evaluate various ML models explored in this work. Nutrition5k is a relatively diverse dataset of mixed food dishes with ingredients level annotations. The dataset contains 20k short videos generated from roughly 5000 unique dishes composed of over 250 different

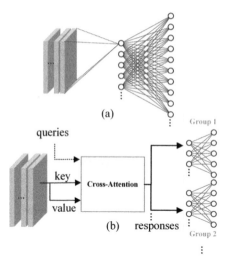

Fig. 3. The decoding mechanisms evaluated in this work: (a) the GAP based decoder averages the latent features before projecting them on to a read-out layer, and (b) the ML-Decoder that computes a response, linear combination of value vectors, against each external query vector. The responses are then projected on to the read-out layer through a group decoding scheme [33].

gle videos, the dataset also exhibits a smaller subset of images collected from a top mounted RGB-D camera, which provides depth images from a top-down view. We constrained our analysis to RGB images only collected from the side-angle video cameras. The data is organized into a 'train' and 'test' subsets, following the original train-test splitting of the video files. The final dataset is obtained by extracting a single image from each video file. To this end, we simply extracted the first frame from each video and downsized it appropriately to be able to process through the neural networks.

This results in an image dataset comprising of around 15K training images and 2.5K test images, all of which are resized to 448x448 resolution in pixels. In Fig. 4, few examples of the dish images from the test split of the dataset are depicted.

ingredients. The dataset also contains the portion estimates of each ingredients, which makes it possible to perform a supervised learning for portion estimation on top of ingredients recognition. The original dataset is collected using video cameras, mounted on the sides of a custom platform to capture each dish from various angles. A digital scale was embedded under the food plates to weigh the dish contents.

The dataset exploits an incremental scanning approach, where a plate is scanned at various time instances with growing cardinality of the ingredients. This resulted in a rich and diverse set of images with varying portion sizes, ingredients, and dish complexity. All incremental scans from a single dish image were organized into a single split to avoid any potential leak of information between the train and test split. Besides side an-

Fig. 4. Examples of dish images from the test dataset.

C. Loss Formulation

Within multi-label learning, a common loss formulation is to consider the final read-out neurons as a series of mutually independent binary classifiers and compute their aggregated binary cross-entropy score. This aggregated score is then used as the minimization objective for the multi-label learning.

Given K labels, each neuron in the read-out layer outputs a score, z^k where $k \in K$, that represents exclusively a single label. Each neuron is then independently activated by a sigmoid function $\sigma\left(z^k\right)$, which converts the logits into probability scores. Let's denote y^k as the ground-truth for the kth class, the total classification loss L^{total} is then obtained by aggregating a binary loss from all labels.

$$L^{total} = \sum_{k=1}^{K} L\left(\sigma\left(z^k\right), y^k\right) \quad (1)$$

A general form of a binary cross entropy loss per label, L, is given by:

$$L = -yL^+ - (1 - y)L^- \quad (2)$$

Where, L^+ *and* L^- are the are the positive and negative parts of the loss respectively, normally evaluated by $L^+ = log(p)$ and $L^- = log(1 - p)$. These parts can be additionally weighted to asymmetrically focus more on the presence or absence of the label. A form of this asymmetric weighing, for multilabel learning, is proposed in [42], which introduces independent focusing parameters, γ^+ and γ- for positive and negative loss parts. This updates the computations as given by (3)

$$L^+ = (1 - p)^{\gamma^+} log(p)$$
$$L^- = p^{\gamma^-} log(1 - p) \quad (3)$$

We used the asymmetric loss, given by (2) and (3), for evaluating all models in this work. The focusing parameters are set to $\gamma^- = 5$ and $\gamma^+ = 0$; which effectively down weighs the loss contribution from easy negatives allowing the network to focus on harder samples as the training progresses.

D. Evaluation Metrics

We employed a mean Average Precision (mAP) metric to evaluate the performance of all ingredient recognition networks, which is a common practice in multi-label learning tasks. The average precision for a single prediction is computed using (4), a micro-average of this is then computed over all the predictions and samples to compute the final score.

$$AP = \sum_{n} (R_n - R_{(n-1)}) P_n \quad (4)$$

Where R_n and P_n are the recall and precisions at the nth thresholds. We used a total of 500 thresholds, equally distributed on the interval [0, 1], to calculate the individual recall and precision scores.

IV. RESULTS

Table I outlines the best mAP scores obtained from the several models evaluated in this work, along with the compute specifications of these models. The compute cost is parameterized by the number of atomic operations and the parameters in the models. The reported mean average precision score is obtained by evaluating the models on the test split of the dataset.

In the first set of experiments, we explored a global average pooling (GAP) based decoding, with different classification neural networks acting as the encoder. In the second set, we selected four encoders to couple with the ML-Decoder, based on their performance in the previous experiments and the compute specifications. The resulting models are then trained with an identical configuration as the previous experiment.

Fig. 5. Prediction results for the dish images selected from the test dataset for illustrative purpose. The detection confidence for each ingredient is shown. The green color represents a true positive detection while the red stands for the false positives.

TABLE I
PERFORMANCE EVALUATION AND BENCHMARK

Models		Performance	Compute	
Encoder	Decoder	mAP %	Operations	Parameters
		%	(GFLOPs)	(MParams)
DenseNet121	GAP	75.6	22.6	7.6
DenseNet169		76.5	26.9	13.6
DenseNet201		74.7	34.3	19.4
MobileNetV1		72.4	4.5	3.8
MobileNetV2		74.5	2.4	2.9
EfficitNetB0		73.3	3.1	4.8
EfficitNetB1		71.9	4.6	7.3
EfficitNetB2		72.2	5.3	8.6
EfficitNetB3		71.5	7.8	11.6
EfficitNetB4		72.7	12.9	18.7
Xception		78.4	36.6	22.0
InceptionV3		72.8	26.4	23.0
MobileNetV2	ML-Decoder	68.0	3.4	9.3
EfficitNetB0		73.4	4.2	11.1
DenseNet169		67.9	28.0	19.9
Xception		70.0	38.0	28.5

Among the models with the GAP decoder, Xception network achieved the best mAP score of 78.4% but is also one of the most computationally expensive networks: it requires 36.6 billion multiply-add operations per image inference. Among smaller networks, the mobileNetV2 is exceptionally performant with 74.5% mAP while using only 2.4 billion operations per inference. Table 1 shows that not all computationally sophisticated models perform equally well. For example, the EfficientNetB1-B4 and the inceptionV3 models did not show

improvements despite their high compute and memory complexity. Upon inspection of the training and validation accuracy curves, we concluded that that these larger models were overfitting the training data and that their performance is capped by the availability of data. However, not all larger models suffer equally from the over-fitting problem. For instance, DenseNet and Xception are comparatively bigger and yet are able generalize quite well on the test dataset. Furthermore, we observed that Efficient-Net models require more careful calibration to reach their full potential. However, to create a fair benchmark, we did not attempt model-specific hyper-parameter tuning sweeps. All networks are trained using a standard training configuration as described in the appendix. All models with GAP decoder scored more than 70% mAP, showing the effectiveness to this simple decoding scheme.

The ML-Decoder did not perform well for the task of ingredients detection. As claimed in [33], the decoder is meant to be a drop-in replacement for GAP-based decoder in a multi-label learning setting. However, our results do not support this claim. Although, some networks perform equally good when coupled with

ML-Decoder, this does not hold in general for all encoder models. For example, when coupled with EfficientNetB0, ML-Decoder performs slightly better compared to using the GAP decoder, while when coupled with other encoders the performance deteriorates. The large performance difference between the models with MobileNetV2 and EfficientNetV2 encoder highlights the weakness of the ML-Decoder, as both encoders are of similar compute sophistication. The root cause analysis of this issues and an ablation study of the decoder is currently considered for future work. Fig. 5. demonstrate few test images, annotated with predictions from a trained model, composed of Denset121encoder and a GAP decoder.

V. CONCLUSION

In this work, we presented a framework for automated diet assessment and demonstrated encouraging results for image-based ingredient recognition using deep learning. The Xception encoder, coupled with a global average-pooling based decoding scheme performs the best, with a mean average precision score of 78.4%. It therefore creates a strong baseline for future work. The attention-based decoding, contrary to what we conjectured, is unable to reliably extract the inter-label relationships and does not improve the overall performance. An ablation study of the attention-based decoder will be attempted further to presumably overcome its current limitations.

APPENDIX

For all training jobs, we used Adam optimizer with an initial learning rate of $1e^{-3}$ following a learning schedule with linear warmup of 200 iterations and a cosine decay to $1e^{-6}$ afterwards. The batch size for all trainings was set to 32 and the models were trained for a maximum of 50 epochs. The encoder of each model is initialized with ImageNet weights. For the ML-Decoder we report outcomes from its default multi-label configuration [33]. Furthermore, we applied a simple data augmentation pipeline, which is composed of random horizonal and vertical flips, random image translations, and random crops with padding.

REFERENCES

[1] Afshin, Ashkan, et al. "Health effects of dietary risks in 195 countries, 1990–2017: a systematic analysis for the Global Burden of Disease Study 2017." The Lancet, vol. 393, no. 10184, pp. 1958–1972, 2019.

[2] Grosso, Giuseppe, et al. "Possible role of diet in cancer: systematic review and multiple meta-analyses of dietary patterns, lifestyle factors, and cancer risk." Nutrition reviews, vol. 75, no. 6, pp. 405–419, 2017.

[3] Feskens, Edith J., et al. "Dietary factors determining diabetes and impaired glucose tolerance: a 20-year follow-up of the Finnish and Dutch cohorts of the Seven Countries Study." Diabetes care, vol. 18, no. 8, pp. 1104–1112, 1995.

[4] Mente, Andrew, et al. "A systematic review of the evidence supporting a causal link between dietary factors and coronary heart disease." Archives of internal medicine, vol. 169, no. 7, pp. 659-669, 2009.

[5] Selhub, Eva. "Nutritional psychiatry: Your brain on food." Harvard and Health Blog, vol. 16, no.11, 2015.

[6] Eva. Keys, Ancel, et al. "The diet and 15-year death rate in the seven countries study." American journal

of epidemiology, vol. 124, no.6, pp. 903–915, 1986.

[7] Sofi, Francesco, et al. "Adherence to Mediterranean diet and health status: meta-analysis." Bmj, vol. 337, Sep. 2008.

[8] Pes, Giovanni Mario, et al. "Evolution of the dietary patterns across nutrition transition in the Sardinian longevity blue zone and association with health indicators in the oldest old." Nutrients, vol. 13, no.5, pp. 1495, 2021

[9] Berry, Sarah E., et al. "Human postprandial responses to food and potential for precision nutrition." Nature medicine, vol. 26, no. 6, pp. 964–973, 2020.

[10] de Toro-Martín, Juan, et al. "Precision nutrition: a review of personalized nutritional approaches for the prevention and management of metabolic syndrome." Nutrients, vol.9, no.8, pp. 913, 2017.

[11] Boeing, H."Nutritional epidemiology: New perspectives for understanding the diet-disease relationship." European journal of clinical nutrition, vol. 67, no.5, 2013.

[12] Thompson, Frances E., et al. "Need for technological innovation in dietary assessment."Journal of the American Dietetic Association, vol 110, no.1, pp. 48, 2010.

[13] Naska, Androniki, Areti Lagiou, and Pagona Lagiou. "Dietary assessment methods in epidemiological research: current state of the art and future prospects." F1000Research, vol. 6, June 2017.

[14] Mimee, Mark, et al. "An ingestible bacterial-electronic system to gastrointestinal health." Science vol. 360, no. 639, pp. 915-918, 2018.

[15] Sahoo, Doyen, et al. "FoodAI: Food image recognition via deep learning for smart food logging." Proceed-ings of the 25th ACM SIGKDD International Conference on Knowledge Discovery & Data Mining, pp. 2260-2268, 2019.

[16] Zuppinger, Claire, et al. "Performance of the Digital Dietary Assessment Tool MyFoodRepo." Nutrients, vol. 14, no. 3, pp. 635, 2022.

[17] Zhang, Zuoyi, et al. "A smart utensil for detecting food pick-up gesture and amount while eating." Proceedings of the 11th Augmented Human International Conference, 2020.

[18] Thames, Quin, et al. "Nutrition5k: Towards automatic nutritional understanding of generic food." Proceedings of the IEEE/CVF Conference on Computer Vision and Pattern Recognition, 2021.

[19] Yanai, Keiji, and Yoshiyuki Kawano. "Food image recognition using deep convolutional network with pre-training and fine-tuning." International Conference on Multimedia & Expo Workshops (ICMEW) IEEE, 2015.

[20] Mezgec, Simon, and Barbara Koroušić Seljak. "NutriNet: a deep learning food and drink image recognition system for dietary assessment." Nutrients vol. 9, no. 7, pp. 657, 2017.

[21] Shroff, Geeta, Asim Smailagic, and Daniel P. Siewiorek. "Wearable context-aware food recognition for calorie monitoring." International symposium on wearable computers. IEEE, 2008.

[22] Herranz, Luis, Shuqiang Jiang, and Ruihan Xu. "Modeling restaurant context for food recognition." IEEE Transactions on Multimedia, vol. 19, no.2, pp. 430-440, 2016.

[23] Ravì, Daniele, Benny Lo, and Guang-Zhong Yang. "Real-time food intake classification and

energy expenditure estimation on a mobile device." IEEE 12th International Conference on Wearable and Implantable Body Sensor Networks, pp. 1–6, 2015.

[24] Matsuda, Yuji, Hajime Hoashi, and Keiji Yanai. "Recognition of multiple-food images by detecting candidate regions." International Conference on Multimedia and Expo. IEEE, pp. 25–30, 2012.

[25] Meyers, Austin, et al. "Im2Calories: towards an automated mobile vision food diary." Proceedings of the IEEE international conference on computer vision, pp. 1233-1241, 2015.

[26] Chen, Liang-Chieh, et al. "Deeplab: Semantic image segmentation with deep convolutional nets, atrous convolution, and fully connected crfs." IEEE transactions on pattern analysis and machine intelligence, vol. 40, no.4, pp. 834-848, 2017.

[27] Min, Weiqing, et al. "Large scale visual food recognition." arXiv preprint arXiv:2103.16107[cs], March 2021.

[28] Ridnik, Tal, et al. "Tresnet: High performance gpu-dedicated architecture." Proceedings of the IEEE/CVF Winter Conference on Applications of Computer Vision, pp. 1400–1409, 2021.

[29] Cheng, Xing, et al. "MlTr: Multi-label Classification with Transformer." arXiv preprint arXiv:2106.06195, 2021.

[30] Liu, Shilong, et al. "Query2label: A simple transformer way to multi-label classification." arXiv preprint arXiv:2107.10834, 2021.

[31] Vaswani, Ashish, et al. "Attention is all you need." Advances in neural information processing systems, vol. 30, 2017.

[32] Zhu, Ke, and Jianxin Wu. "Residual attention: A simple but effective method for multi-label recognition." Proceedings of the IEEE/CVF International Conference on Computer Vision. 2021.

[33] Ridnik, Tal, et al. "Ml-decoder: Scalable and versatile classification head." arXiv preprint arXiv:2111.1293, 2021.

[34] Chen, Zhao-Min, et al. "Multi-label image recognition with graph convolutional networks." Proceedings of the IEEE/CVF conference on computer vision and pattern recognition, 2019

[35] Howard, Andrew G., et al. "Mobilenets: Efficient convolutional neural networks for mobile vision applications." arXiv preprint arXiv:1704.04861, 2017.

[36] Howard, Sandler, Mark, et al. "Mobilenetv2: Inverted residuals and linear bottlenecks." Proceedings of the IEEE conference on computer vision and pattern recognition, pp. 4510–4520, 2018.

[37] Huang, Gao, et al. "Densely connected convolutional networks." Proceedings of the IEEE conference on computer vision and pattern recognition, pp. 4700–4708, 2017.

[38] Szegedy, Christian, et al. "Rethinking the inception architecture for computer vision." Proceedings of the IEEE conference on computer vision and pattern recognition, pp. 2818–2826, 2016.

[39] Tan, Mingxing, and Quoc Le. "Efficientnet: Rethinking model scaling for convolutional neural networks." International conference on machine learning. PMLR, pp. 6105–6114, 2019.

[40] Chollet, François. "Xception: Deep learning with depthwise separable convolutions." Proceedings of the

IEEE conference on computer vision and pattern recognition, pp. 1251–1258, 2017.

[41] Chollet Russakovsky, Olga, et al. "Imagenet large scale visual recognition challenge."International journal of computer vision, vol. 115, no.3, pp. 211–252, 2015.

[42] Ridnik, Tal, et al. "Asymmetric loss for multi-label classification." Proceedings of the IEEE/CVF International Conference on Computer Vision, 2021.

Rameez Ismail received master in robotics (2015) from Technical Univ. Dortmund, Germany, and Engineering Doctorate in systems design (2017) from Eindhoven Univ. of Technology, Netherlands. He started his career with a brief stint (2016-2018) at NXP Semiconductors and thereafter joined Philips Research as a scientist with the ambition to accelerate the digital transformation of healthcare. His research interests include Artificial Intelligence (AI), biomimetics systems and High-Performance Computing (HPC). Over the years, he successfully applied advanced AI research to various healthcare use cases within the scope of personal health and medical imaging.

Index